columbus

IN SPACE

Europe's voyage of discovery
on the International Space Station

European Space Agency

1 3 5 7 9 10 8 6 4 2

Century
20 Vauxhall Bridge Road
London SW1V 2SA

Century is part of Penguin Random House group of companies whose
addresses can be found at global.penguinrandomhouse.com.

Penguin
Random House
UK

Written by: Julien Harrod
Illustration and design: Attilio Brancaccio
Images: © ESA / NASA

First published in 2018 by Century
www.penguin.co.uk

A CIP catalogue record for this book is available from the British Library.
Trade paperback **ISBN 9781780899312**

Printed and bound in India by Replika Press Pvt. Ltd.

MIX
Paper from
responsible sources
FSC® C018179

Penguin Random House is committed to a sustainable future for our
business, our readers and our planet. This book is made from Forest
Stewardship Council® certified paper

CONTENTS

Any chance to fly in space is a rare privilege. But it was even more special to be a part of the crew that delivered the Columbus laboratory to the International Space Station. Our Space Shuttle mission, STS-122, was the 29th flight of Atlantis and the 121st of the Shuttle program – just one of many. But in the Space Station assembly sequence, our flight was designated '1E', with the 'E' standing for 'Europe'. No other mission had that letter in its name; ours was the first and last. Although our spacecraft and most of our crew members were American, our mission patch depicted a Renaissance-era sailing ship to highlight the link between European voyages of discovery centuries ago and today's first exploratory steps into the vast universe that begins at the top of our atmosphere. It was a proud moment for my crewmates and me when we undocked Atlantis from the International Space Station, gently pulled away, and saw the full span of the Station with the brand-new Columbus module in its rightful spot, on a berthing port that had been empty when we arrived. We felt like we'd added important new capabilities and resources to the Station, and done our part to establish Europe's place in the new world of space.

Stan Love
NASA Astronaut

*"In fourteen hundred and ninety-two
Columbus sailed the ocean blue…"*

In 2008 Columbus launched into the blackness of space.

This book is about a laboratory. Can you name any laboratory on Earth? Maybe not, so you might be wondering why this book. Well, Columbus is a 7-m long, 10-tonne laboratory – not on Earth but circling our planet at a speed of 28,800 km/h, and it has been doing so for over 10 years.

Space is a tough place to work, let alone build a laboratory. There is no oxygen to breathe, and every 90 minutes the Columbus laboratory and the International Space Station do a complete orbit of Earth – going from sunlight to darkness and back again – with temperatures fluctuating by up to 300°C each time it passes from sunlight into Earth's shadow. We haven't even mentioned getting to space in the first place.

How the European Space Agency conceived, launched, installed and operates its space laboratory is a part of this book, but it also details the daily work happening 400 kilometres above our heads and its importance. Fundamental research that can only be performed in space is taking place as you read this sentence, and a mass of information is piped through satellites to the Columbus Control Centre in Germany for processing 24 hours a day.

The captivating space race of the 1960s culminated in the first steps of human beings on the Moon. Now spaceflight has become so common that an average of seven rockets a month were launched across the globe in 2017. From your smartphone to your holiday flight, almost everything you do communicates through satellites and uses space technology.

Anyone born after November 2000 has never lived without a human being orbiting our planet. The space age is here, and we are living in it. The Columbus laboratory is Europe's space in space, exploring our Universe to benefit us all.

COLUMBUS FROM VISION TO MISSION

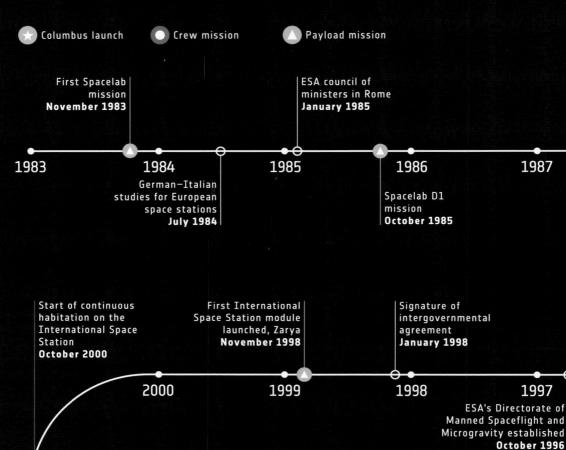

★ Columbus launch ● Crew mission ▲ Payload mission

First Spacelab mission
November 1983

ESA council of ministers in Rome
January 1985

1983   **1984**   **1985**   **1986**   **1987**

German–Italian studies for European space stations
July 1984

Spacelab D1 mission
October 1985

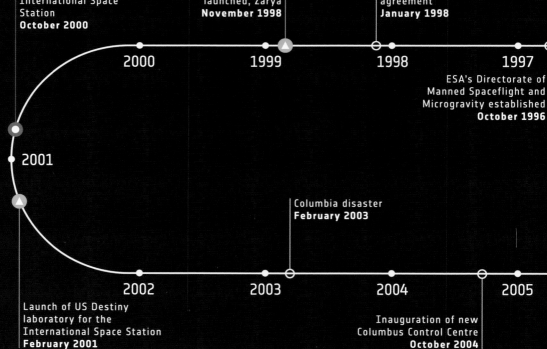

Start of continuous habitation on the International Space Station
October 2000

First International Space Station module launched, Zarya
November 1998

Signature of intergovernmental agreement
January 1998

2000   **1999**   **1998**   **1997**

ESA's Directorate of Manned Spaceflight and Microgravity established
October 1996

2001

Columbia disaster
February 2003

2002   **2003**   **2004**   **2005**

Launch of US Destiny laboratory for the International Space Station
February 2001

Inauguration of new Columbus Control Centre
October 2004

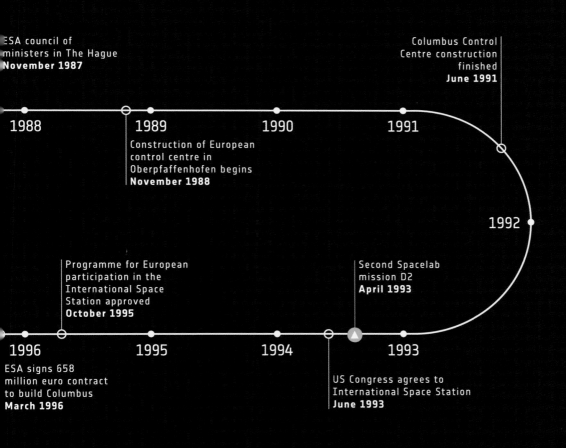

ESA council of
ministers in The Hague
November 1987

Columbus Control
Centre construction
finished
June 1991

1988   **1989**   **1990**   **1991**

Construction of European
control centre in
Oberpfaffenhofen begins
November 1988

1992

Programme for European
participation in the
International Space
Station approved
October 1995

Second Spacelab
mission D2
April 1993

1996   **1995**   **1994**   **1993**

ESA signs 658
million euro contract
to build Columbus
March 1996

US Congress agrees to
International Space Station
June 1993

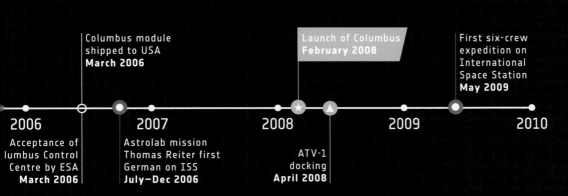

Columbus module
shipped to USA
March 2006

Launch of Columbus
February 2008

First six-crew
expedition on
International
Space Station
May 2009

2006   **2007**   **2008**   **2009**   **2010**

Acceptance of
Columbus Control
Centre by ESA
March 2006

Astrolab mission
Thomas Reiter first
German on ISS
July–Dec 2006

ATV-1
docking
April 2008

LAUNCHING COLUMBUS

From Spacelab to Space Station

Like Christopher Columbus's transatlantic voyage to discover an unknown world in 1492, the European Columbus laboratory was meticulously planned, budgeted, scrapped and redesigned before getting the official blessing to build the ship before launch. Finally, with support from all over Europe and international partners, Columbus set sail on a voyage for humanity's new world in space on 7 February 2008.

The origins of Columbus can be traced to Europe's first orbital laboratory — Spacelab, which first launched on Space Shuttle Columbia in 1983. Unlike the Columbus laboratory, Spacelab was not attached to a space station but flew on 22 missions inside the US Space Shuttle, the longest mission lasting 17 days.

Designed to fit snugly in the Space Shuttle's cargo bay, Spacelab was a suite of elements that could be mixed and matched to serve each individual mission's purpose. Elements included an internal area where astronauts could work as well as exterior pallets so experiments could be exposed to space.

Around the time of Spacelab's first launch, a feasibility study into a more permanent space lab took place. From this came the idea for Columbus — in a time without internet, laptop computers, touch screens or rapid genetic sequencing. (Thirty years later these technologies are used daily by astronauts in Columbus.)

The multi-purpose design of Spacelab greatly influenced the design of Columbus. With only one laboratory to work with and the whole universe of possibilities that spaceflight offered, Columbus was designed to accommodate as many disciplines as possible in a small place. The new laboratory was designed to be equipped with facilities to conduct science in biology, physiology and fluids with further plans for a muscular-research machine and a self-contained glovebox for working on dangerous experiments with chemicals and fire. As with Spacelab, external facilities were not neglected either. A Sun monitor and a facility to expose samples to space were included, with more planned to be installed later.

From astrobiology to solar science through metallurgy and psychology — researchers from across the spectrum of scientific fields were to be able to run experiments on Columbus to investigate the unknown and apply the knowledge to master phenomena on Earth.

Building the Module

After an agreement was signed to participate in the International Space Station with Russia, Japan, Canada and the United States, contracts to build Columbus were put in place in 1996. Two years later, its design review was completed and the vision began to take shape. While engineering teams worked on the hardware of the space-bound laboratory, others thought of how to control it from Earth. In 1998 it was decided to build the Columbus Control Centre at the German Aerospace Center near Munich.

Each component that flies into space has to endure the harsh conditions outside our atmosphere as well as the intense vibrations and forces of a rocket launch. Every part of Columbus had to be designed and built to withstand eight minutes of intense acceleration followed by many years operating in weightlessness – and work flawlessly. Space designers, engineers and technicians were constructing a laboratory that would survive and work in opposites – the rough vibrations of launch followed by the relative calm of coasting in weightlessness.

By 2000, a full-scale model was delivered to NASA's testing facility at the Johnson Space Center in Houston, USA. The astronauts that would install Columbus could now practise the spacewalks that would connect the living-room-sized module to the International Space Station. Meanwhile extensive computer tests took place to make sure Columbus communicated well with its sister modules. A year later, most safety reviews and tests had been passed, and a launch date was tentatively scheduled for 2004.

From Turin, Italy, the flight model (the version of Columbus that would actually go to space) travelled to Bremen, Germany, for final assembly. By the end of 2002, most components were in place and wired in, followed by the first system test of the final product.

A second full-size Columbus model was also made and delivered to the European Astronaut Centre in Cologne, Germany, to train astronauts on the systems inside. Training in the model is mandatory for all astronauts that visit the International Space Station. ESA's technical heart in the Netherlands has a third full-scale model of Columbus on display to help researchers design experiments.

Qualification tests for Columbus were completed by 2003, and the module was branded fit-to-fly. To ensure nothing would fail, every element of the flight model was tested to extremes by recreating space on Earth and pushing the hardware to its limits. Vacuum tests, repeated temperature cycles, solar radiation and emergency scenarios were all part of this ultra-stringent quality control.

Unfortunately, it was around this time that launch plans came to an unexpected stop due to unforeseen circumstances.

On 1 February 2003, Space Shuttle Columbia, the same spacecraft that flew the first Spacelab mission, disintegrated on re-entry into Earth, killing all seven astronauts inside. This tragic loss put an indefinite hold on the Space Shuttle programme as investigations were held to find out what went wrong and improve safety before another launch. The Columbus payload was forced into hibernation. It wouldn't be until 2008, some five years later, that Columbus finally got off the ground.

Columbus was designed and built with new technologies that had never been produced in Europe before. Nine countries agreed on the scope and worked together to produce this technical feat within budget (900 million euros) and further negotiated, through the European Space Agency, its inclusion on the International Space Station – bartering with the USA, Russia, Japan and Canada.

Many aspects of the project were offered and exchanged in kind. It was NASA who eventually flew Columbus into space on its Space Shuttle Atlantis in exchange for a share of its use, with ESA further contributing to the running costs of the International Space Station through its Automated Transfer Vehicle supply-and-support spacecraft.

From idea to reality in less than 30 years, the building of the Columbus space laboratory is a shining example of fine planning, efficient collaboration and excellent negotiation by all involved. The next step, however, was to prove more arduous: shipping the laboratory to space and installing it on the International Space Station – all 2.5 million moving parts of it.

A word on gravity

The exact definition of where space starts varies but is generally considered around 100 kilometres above sea level. Getting anywhere near that high requires escaping the pull of Earth's gravity – and burning a lot of fuel in doing so.

Contrary to what you might think, a rocket does not launch straight up, but turns to fly perpendicular to Earth as it gathers enough speed to stop falling back.

It was the first person to write down the mathematics behind gravity, Sir Isaac Newton himself, who first considered what would happen if a cannon ball was fired straight towards the horizon. This was in 1687 before rocket engines, jet engines or even steam engines were invented — aside from horses, the cannon ball was the fastest thing controlled by humans at the time.

Common assumption at the time dictated that the ball would fall back to Earth once its speed — and thus its trajectory — decreased. More explosives packed into the cannon would make the ball fire faster and travel further before gravity pulled it back to the ground. What would happen, Newton pondered, if you fired the cannon ball so fast that by the time gravity pulled it down it had travelled far enough to avoid hitting Earth altogether? As a good scientist should, Newton provided a hypothesis to his own question, stipulating that as long as enough speed was maintained the cannon ball would continue circling Earth indefinitely, in effect reaching orbit.

When Newton published this idea in his book Philosophiæ Naturalis Principia Mathematica, it was but a hypothetical — a thought-experiment outside the realm of possibility. By the 20th century, however, human invention had progressed beyond Newton's idea and humans were getting so good at experimentation that we were launching space stations — and the Columbus module.

What speed did the cannon ball need to travel to reach orbit? Each planetary body has its own orbital velocity or the minimum speed required to reach its orbit. Earth's orbital velocity is just over 28,440 km/h at the equator at sea level. Furthermore, the higher the orbit, the faster an object needs to travel. The International Space Station's cruising speed is therefore 28,800 km/h. Columbus would need to beat that speed to rendezvous with the Station and be installed, all while circling 400 kilometres above Earth.

Catching the International Space Station

To get Columbus into low Earth-orbit and catch up with the International Space Station, it needed be launched to the correct altitude and speed.

The vehicle chosen for this task was NASA's Space Shuttle Atlantis. Mission STS-122 would

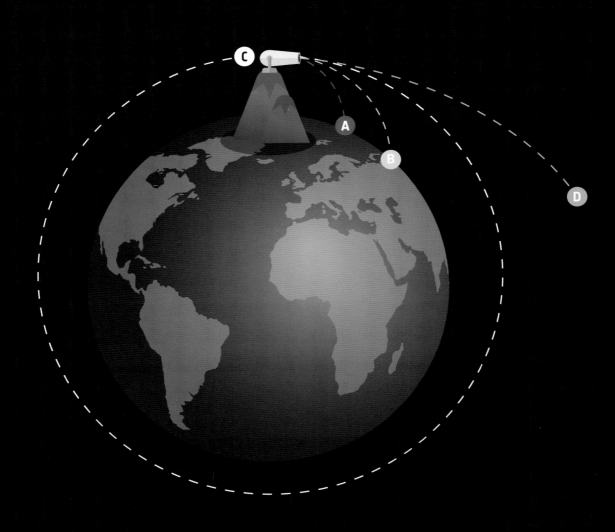

Orbital velocity ground level:
28,440 km/h
Low Earth-orbit:
23,222 km/h – 28,083 km/h
International Space Station:
28,800 km/h

Too slow,
speed is lower than
gravity's pull

Getting there...

Escape velocity:
40,269.6 km/h

send NASA astronauts Stephen Frick, Alan Poindexter, Leland Melvin, Rex Walheim and Stanley Love and European Space Agency astronauts Hans Schlegel and Léopold Eyharts together with the Columbus module into space. Aside from the Columbus module itself, the cargo roster included the science facilities for inside the European space lab, two experiment facilities that would be installed on the outside, a Canadian-built robotic arm, fuel and a number of supplies for the crew to eat, drink and breathe during their mission.

The total weight of supplies — including Columbus — for the STS-122 mission was 17,311 kilogrammes, though the Shuttle and its expendable rockets brought the total weight up to 2 million kilogrammes — or 2,000 tonnes. The vast majority of this weight was the fuel that would be ignited and burnt at alarming rates — astronauts nervously liken lift-offs to controlled explosions.

The huge difference between total weight and the material being sent into space (called a 'payload' in space jargon) is one of the reasons rocket launches are relatively rare and expensive. You need a lot of energy to get into space, which means burning a lot of fuel extremely quickly — almost half a million kilogrammes (equivalent in weight to around 100 elephants) of solid propellant is burnt in 480 seconds.

Spending fuel

On 7 February 2008, launch day for Columbus, technicians at the launchpad at Kennedy Space Center in Florida, USA, began pumping fuel in the early hours of the morning — it takes quite some time to pump two million litres of liquid oxygen and liquid hydrogen into a rocket.

The seven astronauts sitting in the cockpit of the Shuttle had a tense wait of over two hours as ground technicians prepared everything for blast-off. Launch had already been delayed by a few months due to fuel-pump problems. The last thing the astronauts wanted was the launch to be scrapped and put off again, and the weather in February was far from ideal.

Like catching a bus to be on time at a meeting, you cannot just launch a rocket into space at any time without considering the final destination, and whether you'll make it on time. A Space Shuttle mission to the International Space Station has only five minutes a day during which it can be launched; failure to take off inside the time when the orbits of Earth and the outpost are aligned would mean that the spacecraft could not catch up with its destination and the launch would be put off to another day.

As late as 90 minutes before the launch window opened, the NASA meteorologist declared a "No Go for launch" – a worrying thunderstorm was developing to the west of the Kennedy launch centre in Florida. Thankfully, half an hour before the launch window the skies cleared: the mission was going to happen!

At 19:45 GMT the countdown ended, engines ignited and Space Shuttle Atlantis rumbled towards the sky. After years of preparations, many things happened in quick succession.

Fifty seconds before launch the Shuttle was disconnected from its launch tower to run on its own power and take over the countdown. Twenty-one seconds before launch the Shuttle's engine gimbals were tested by moving them along all axes.

With 16 seconds to go the launch pad was flooded with 113,000 litres of water pumped through two pipes large enough to walk through. The water absorbs vibrations and sound to protect the support structure and spacecraft itself as well as forming the iconic plumes that bloom during lift-off – not smoke but water vapour!

On average for every second during launch, the astronauts and Columbus were accelerated by 100 km/h. Two seconds after launch they were travelling 200 km/h, a second later 300 km/h and so on, not stopping until they reached orbit eight minutes later.

It took just 35 seconds after lift-off for Atlantis to break the sound barrier, travelling at speeds over Mach 1, and it reached Mach 23 at the end of its acceleration. The Space Shuttle's thrust is so impressive that soon after lift-off Atlantis's main engine actually had to throttle back to avoid accelerating too much.

During the eight-minute acceleration period everything and everybody strapped inside and to Atlantis experienced 3g force – so three times the force you feel normally of Earth's gravity pulling you down. If you want to get a rough idea of how this feels, ask two people who weigh the same as you to lie on top of you for eight minutes while you continue to work – uncomfortable, yes, but nothing life-threatening.

The Columbus laboratory secured inside Atlantis' payload bay also experienced 3g force – weighing 38,325 kilogrammes instead its normal 12,775 kilogrammes for the eight-minute blast-off.

SHUTTLE LAUNCH PAD

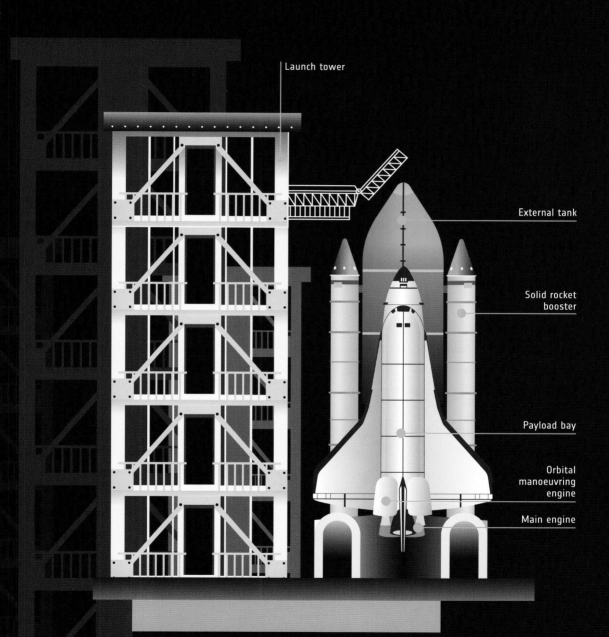

Launch tower

External tank

Solid rocket booster

Payload bay

Orbital manoeuvring engine

Main engine

Two minutes after launch the solid rocket boosters had spent all their fuel and were jettisoned into the ocean, to be retrieved and refurbished for reuse on another mission.

As Atlantis continued upwards its last remaining external tank still pumped fuel until 19:54 GMT when the main engine was switched off. The mostly empty, large orange fuel tank was left to fall back to Earth.

From ignition, the Space Shuttle and its rocket boosters burnt all their fuel in just eight minutes — keeping a maximum reserve of 13,000 litres to manoeuvre in space and return to Earth.

Atlantis was now flying at around 28,800 km/h and after a quick engine burn was in an orbit chasing the International Space Station. After the intense sounds and vibrations of launch, the astronauts inside were officially in space and could float freely in the quiet hum of the Space Shuttle.

Orientation

The next two days, 8 and 9 February, were spent catching up with the International Space Station. All was looking good for the intricately designed space equipment on board, with Columbus being kept nice and warm; the humans inside the Space Shuttle, however, were not designed to leave Earth and needed to quickly adjust to life in space.

Of the seven astronauts in Atlantis, three — Alan Poindexter, Leland Melvin and Stan Love — were on their first spaceflight, with Stephen Frick, Rex Walheim, Hans Schlegel and Léo Eyharts considered veterans having already flown in space at least once before.

Regardless of experience, adapting to weightless life takes some time. Humans have evolved on planet Earth to function with the continuous pull of 9.81 m/s^2 towards Earth's centre of gravity. Take this away suddenly and the body goes slightly haywire.

Columbus research published nine years after the Atlantis launch revealed that human cells go into disarray but recover fully in exactly 42 seconds after exposure to weightlessness. For those inside Atlantis, this knowledge would have been little comfort as fluids in their bodies rushed to their heads and their backbones expanded. A human heart is made to pump blood that naturally pools towards our feet when standing or sitting. In space this downward tug no longer applies and our hearts have to adapt to the new environment —

countering what millions of years of evolution have ensured they do best.

Similarly, sinuses congest as fluids rise to the head, leading to what astronauts lovingly call "puffy-head, chicken-leg syndrome". Pictures of astronauts in space and on Earth show that living in weightlessness gives you a fuller, plumper face. Some astronauts have even commented that living in space is like living with the symptoms of a permanent cold.

The extra fluid that collects around the brain also influences the head and is partly the reason for almost three-quarters of astronauts reporting they suffer from headaches in space. Described by some as 'exploding', the headaches are unlike anything felt on Earth. Surveys developed by European scientists for astronauts who visit the Space Station are showing that these new types of headaches require an entirely separate classification to any of those suffered on Earth.

Back pain is another common astronaut ailment, as their spines expand without the usual forces of gravity to keep the vertebrae together. After a night's sleep on Earth, your spine's natural stretch will mean you wake up over a centimetre taller than you were the night before. During the day as you go about your business, sitting and walking, the discs in your spine compress and it becomes shorter again. For most people this is no issue as the body is used to it, but in space, where there is no walking or sitting, just floating, the spine stays elongated and some astronauts grow by up to seven centimetres during spaceflight – with back pain as a side effect.

The brain also needs to cope with many new and unnatural signals. On Earth, fluid in our inner ear helps us keep a sense of balance by acting as a spirit level. Every kid has turned in circles quickly to make the world spin even when they stop turning – the spinning is caused by the liquid in the inner ear sloshing around with the brain interpreting the sloshing as if the body is still moving. Imagine how the brain copes when astronauts are weightless in space and the fluid in their ears floats and sloshes as never before. ESA astronaut Tim Peake did a quick test at the end of his mission to determine if he could still get dizzy after six months in space. He asked his crewmate Tim Kopra to spin him around for a few seconds and then stop him spinning, just like a kid would do on Earth. The conclusion: no dizziness, as the brain had learnt to interpret these signals differently.

So how do astronauts orient themselves in space? The simple answer is nobody really knows for sure right now, but the question is a focus of a large part of the studies that are conducted on Columbus. Considering that there is no clear up or down in space and no

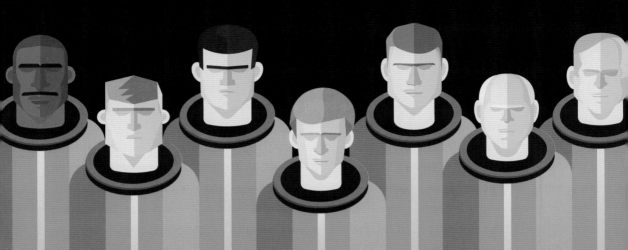

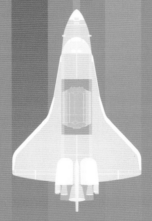

Melvin | Walheim | Love | Schlegel
Frick | Eyharts | Poindexter

sts-122

ATLANTIS

horizon to orient yourself, the human body adapts remarkably well. Within hours, the brain adjusts to the lack of normal sensory input and within days, astronauts float majestically from module to module as second nature. Handy, because after two days chasing the International Space Station, the crew in Atlantis were getting ready to connect two multi-billion-dollar spacecraft while flying at 23 times the speed of sound.

Rendezvous and contact

February 9, the third day for the first European Space Station construction mission, had the crew of Atlantis preparing for docking with the orbital outpost where Columbus would stay for the rest of its working life — the International Space Station.

The astronauts spent most of the day inspecting the Shuttle orbiter's shields that protected it from the intense heat that builds up when entering Earth's atmosphere. After the 2003 Columbia Space Shuttle re-entry disaster, NASA investigations concluded that foam insulation had got loose during launch and damaged the ceramic tiles that protect the Shuttle on re-entry. A video camera on the end of the Shuttle's robotic arm inspected the protective insulation as part of the improved safety routine for each mission.

Mission control analysed the video and reported everything was still as it should have been. They also confirmed there was enough fuel and supplies to extend the Columbus installation mission by an extra day, allowing some extra time for the astronauts and mission control to address any issues that came up.

The rendezvous and docking sequence would begin from a point some 70 kilometres behind the International Space Station, at the same altitude. At 13:06 GMT Alan Poindexter piloted Atlantis to a lower orbit, thus gaining speed and overtaking its target.

Many things need to be calculated and planned before two spacecraft can connect. The International Space Station first moved its fragile solar panels to protect them from the Shuttle engines. As the solar panels now generated less electricity, electrical systems were switched off in the outpost.

The planners timed the docking to get the best lighting conditions, not for pretty pictures but because nobody wanted to dock the most expensive structure in the world blinded by a glaring sunrise.

The Space Station's guidance and orientation system was then turned off to no longer keep it pointing in its normal position. If the system was left on it would try to counteract the bump of the Shuttle as it docked – causing huge stress on the system and on the Station's hull. In effect, before docking, the Space Station was turned into a passive hunk of metal drifting freely over our planet.

After a 90-minute orbit of earth, the Space Shuttle returned to roughly the same altitude as the International Space Station. It was now only 12 kilometres behind the outpost. The final approach included a majestic tumbling somersault to show the orbiter's belly to the crew on the Space Station. Inside the larger outpost, cosmonaut Yuri Malenchenko had lived and worked with NASA astronauts Peggy Whitson and Dan Tani for three months as the only three humans living off Earth – they were eagerly awaiting the arrival of new faces.

With the loop-the-loop completed Atlantis moved to a few hundred metres in front of the Space Station and aligned its docking port with the Harmony module on the International Space Station.

At 50 metres in front of the Station, Atlantis received a "Go" for docking from mission control and Alan started the approach. Relative to each other the two objects in space were moving around two metres a minute as the 90-tonne Space Shuttle edged towards the 408-tonne Space Station, but they were still travelling at around 28,800 km/h over Earth – a good thing or they would slowly start falling back to our planet.

A Shuttle docking to the International Space Station has a 7.62 cm margin of error. Alan spent many hours training on Earth finely rehearsing and here he was aided by lasers, cameras and his crewmate's instruction on the radio. With so much at stake and such small margins between success and disaster, the docking between two spacecraft is a tense moment in all human spaceflight missions.

At 17:17 GMT, while coasting over Australia, the two spacecraft docked. Alan called over the radio, "Houston, Alpha and Atlantis. We have capture confirmed." Columbus had reached its home for the next decade and more.

Before the new arrivals could greet the astronauts on the other side of the hatch, extensive tests were done. Airtight connections are hard to achieve on Earth so when your life depends on limited amounts of air while surrounded by the vacuum of space, a few leak

SPACE SHUTTLE DOCKING WITH INTERNATIONAL SPACE STATION

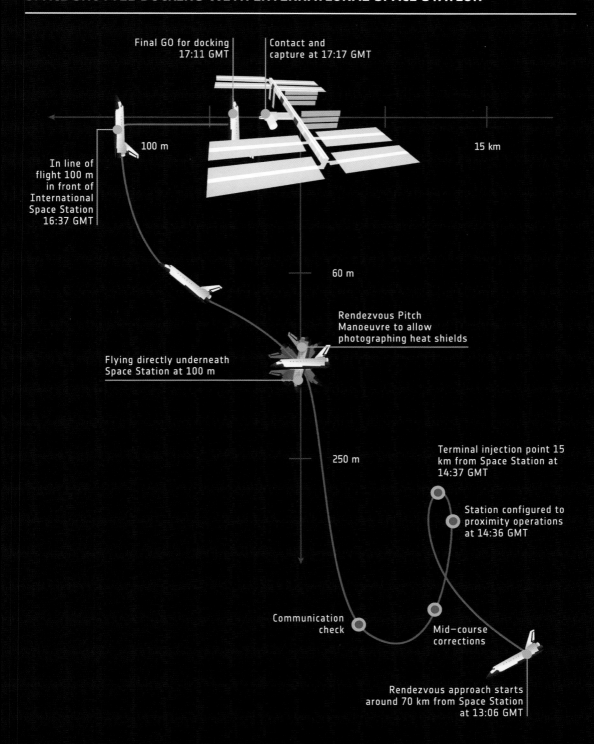

Final GO for docking
17:11 GMT

Contact and
capture at 17:17 GMT

100 m

15 km

In line of
flight 100 m
in front of
International
Space Station
16:37 GMT

60 m

Rendezvous Pitch
Manoeuvre to allow
photographing heat shields

Flying directly underneath
Space Station at 100 m

Terminal injection point 15
km from Space Station at
14:37 GMT

250 m

Station configured to
proximity operations
at 14:36 GMT

Communication
check

Mid-course
corrections

Rendezvous approach starts
around 70 km from Space Station
at 13:06 GMT

tests are recommended before opening any hatch. The crew tested the connections by pumping extra air and monitoring the pressure — if the pressure held, there were no leaks. A full orbit of Earth later and the pressure remained. The hatches were opened and the seven astronauts in Atlantis greeted their three colleagues in the International Space Station who had been living isolated in space for months.

INSTALLING A SPACE LAB

Slight change of plans

On the evening after docking, on 9 February, NASA announced a reschedule of the first spacewalk to install Columbus, moving it back by a day as well as changing who would conduct the spacewalk. Hans Schlegel was not feeling well and would not head outside on the first spacewalk — Stan Love would take his place. Though Stan had trained for hours on Earth, an extra day of revision always improves results, be it for an exam, an important deadline or an unexpected spacewalk.

The world's largest indoor swimming pool is at NASA's Johnson Space Center in Houston, Texas, where astronauts train underwater in their full space suits and gear on a full-size model of the International Space Station. Fitted with weights, the astronauts can achieve neutral buoyancy and float in the water similar to when on a spacewalk 400 kilometres above Earth — just 12 metres underwater instead.

Aside from the comforting knowledge that life-sustaining air is only a few metres above, the main difference between training underwater and the real thing in space is drag. An object in space is easy to move and difficult to keep still; moving something underwater is difficult as the water causes drag and objects will quickly stop. Consider stirring an object first in honey, then in oil, water, air and finally in a vacuum. Each environment will be progressively easier to move in, with vacuum being the easiest as there is no resistance or drag from atoms to push against.

Nevertheless, diving underwater is the next best thing to practise the extremely complex routines a spacewalk requires. Each minute of a spacewalk is planned and choreographed to be as efficient and safe as possible. Tool belts are prepared in advance and tools are clipped in by order they are required — every hand movement is considered in advance, a meticulous choreography in the heavens.

The International Space Station is a large laboratory measuring 102 metres long – making it larger than a football field. Many areas on the Station are fragile or simply dangerous to touch, so for safety a spacewalk is always a collaboration between two astronauts working with the 'buddy system', each keeping an eye on themselves and their partner. Support comes from inside the Space Station and ground control with all teams working in unison. A plan for a typical spacewalk totals over 50 pages of specific information, acronyms and warnings that are memorised by the duo heading outside.

Once a spacewalk is decided, mission control plans as many tasks as possible, so spacewalks often last over six hours. During this time the spacewalkers cannot eat (they have a straw connected to a water pack) or go to the toilet (they wear high-tech nappies) or scratch their nose (they have a pad in the helmet they can rub against). Even a sneeze can have serious consequences – if it dirties the inside of the helmet, there is no way to clean the visor.

Stan had a lot to consider as he and Rex Walheim prepared for the next day's sortie. On the inside, Léo Eyharts would be operating the Station's robotic arm to move Columbus into position, and he too needed to be in sync with the duo outside. A six-hour dance in the heavens had been prepared for months, and this was the last chance for revision before the big show.

Going outside to connect Columbus

Stan and Rex slept in the Space Station's airlock to prepare for the spacewalk. One of the many things a spacesuit is designed for is protection from the vacuum of space. Humans need a minimum air pressure to breathe and function, but for a spacesuit designer the lower the pressure inside the spacesuit the easier, as it reduces the difference between suit and outside pressure. The smaller the pressure difference, the easier it will be to move the spacesuit, similar to how an overinflated bicycle tyre will feel stiff and not move to the touch, while a flat tyre with little air is supple and can be easily compressed by hand.

So spacesuits are designed to operate at less pressure. The human body can handle changes in ambient pressure pretty well, allowing us to explore mountain ranges as well as dive underwater to investigate the seas. Rapid pressure change can be problematic and

COLUMBUS CONTROL CENTRE

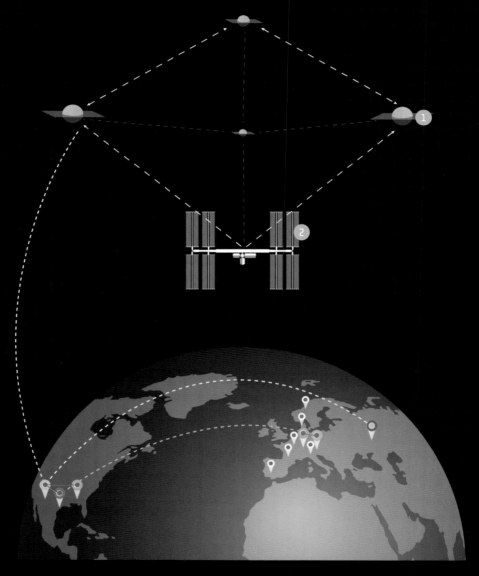

White Sands
New Mexico, USA

Space Station Mission
Control Center
Texas, USA

Huntsville Operations
Support Center
Alabama, USA

Columbus
Control Centre
Oberpfaffenhofen, Germany

European Astronaut Centre
Cologne, Germany

User Support and
Operations Centres

Russian Mission
Control Centre
Moscow, Russia

1 Satellite
network

2 International
Space Station

cause what divers call the "bends". Naturally occurring nitrogen in our blood can inflate into small bubbles as pressure around us decreases. The bubbles can occur anywhere in our body and are invariably harmful, causing rashes, joint pain, brain aneurisms and even death.

Taking it slow and adapting to a change in pressure allows nitrogen to dissolve before it bubbles. A second way to avoid the bends is to breathe pure oxygen and purge nitrogen from the body. Rex and Stan would be using both techniques on their spacewalk: after spending the night in the Station's airlock at reduced pressure, they would breathe pure oxygen while out on the spacewalk.

In the morning of 11 February, Stan and Rex awoke and repressurised the airlock briefly so they could brush their teeth and go to the toilet. During their morning rituals they wore oxygen masks to keep the nitrogen levels in their bodies low.

Peggy Whitson and Alan Poindexter helped the pair into the suits that would keep them safe for the next eight hours. Aside from the pressure, the suits needed to insulate them from temperature differences of 300°C and provide water. Standard spacewalk procedure is like mountaineering: at least one tether should be connected at all times to secure you if you lose grip. For safety the suits come equipped with a last-resort jetpack that could propel spacewalkers back to the International Space Station if they lost their grip (the jetpack has never been used outside of testing).

Rex and Stan opened the outside airlock and entered the void of space at 14:14 GMT to make their way to the Harmony module where Atlantis was docked – Columbus would also berth to Harmony but on another port. Rex first went to Columbus's future port and removed its protective cover while inspecting the sealing rings for any problems. He then joined Stan, who was preparing the Station's robotic arm.

Stan was going to strap his feet to the end of the 16-m robotic arm for the ride of a lifetime. Inside the Space Station, Alan was set to control the arm and move Stan around.

To prepare for the move of Columbus it needed to be unplugged from the Space Shuttle's systems, which had been keeping it at a minimum temperature. When Alan switched off the power to Columbus he started a countdown clock. The laboratory would be getting colder and colder until power returned – too long and equipment could break or cooling water freeze and burst pipes. Nobody knew exactly how long Columbus could survive unpowered but best estimates set the countdown to 19 hours maximum. More than enough time if everything proceeded as planned.

Once the switch was flipped, Alan floated inside Columbus from the Space Shuttle and physically disconnected the power cable. The robotic arm now had to connect with Columbus and attach to a grappling point, which would allow power to pass through. Once the robotic arm made contact the heaters inside the lab would be repowered.

The grappling port on Columbus wasn't ready yet, though. It had been too large to fit in the Shuttle payload bay, so Stan was swung over to Atlantis and back to collect it.

By the time Stan got to where the port needed to be installed, Rex had reached the same destination the old-fashioned way, clambering along the Space Station and Space Shuttle using his hands.

By 19:36 GMT they finished installation of the grapple port and Stan released himself from the robotic arm to help connect Columbus. Four hours after disconnection, power was restored to Columbus. Now it just needed to be secured in place.

Rex moved over to the future berthing point for Columbus on the starboard side of Harmony, and removed its protective cover.

At 19:53 GMT the flight director at NASA's mission control gave the go-ahead to unberth Columbus and start it on its short but steady journey to Harmony. Under Alan's command, the 12.8-tonne module inched out of the Atlantis payload bay.

At 21:29 GMT contact was made between Columbus and the International Space Station, and physical installation was complete. Since that moment, Columbus has not moved from its berthing space — though, of course, it has now orbited earth over 58,000 times (and counting)!

With the berthing of Columbus, European mission control in Munich, Germany, now had its work cut out to set it up. The first command sent to Columbus returned a failure report, but was met with great enthusiasm by everyone: it proved that the elaborate ground-to-space network was running. At that time Columbus was simply a heated module, but the fact that it could return an error code when questioned was an achievement in itself. Unfortunately, it would not be the last error received.

INTERNATIONAL SPACE STATION

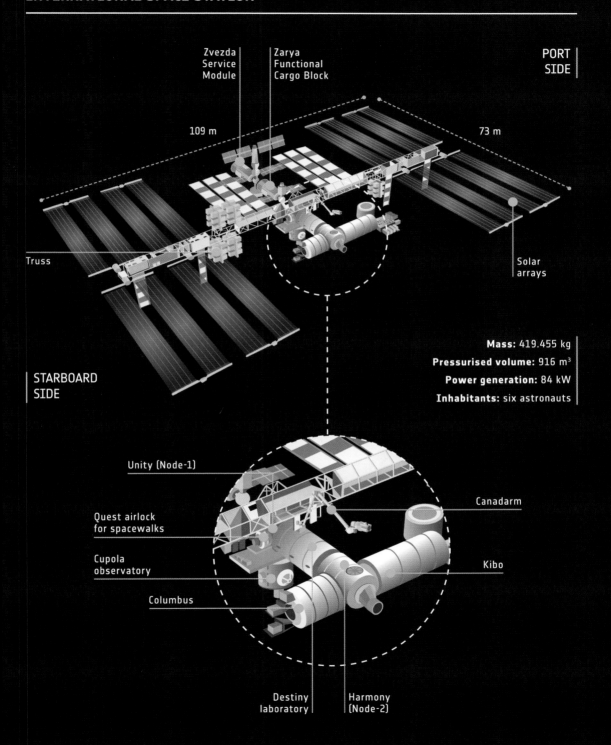

Zvezda Service Module

Zarya Functional Cargo Block

109 m

73 m

Truss

Solar arrays

STARBOARD SIDE

Mass: 419.455 kg
Pressurised volume: 916 m³
Power generation: 84 kW
Inhabitants: six astronauts

Unity (Node-1)

Quest airlock for spacewalks

Cupola observatory

Columbus

Canadarm

Kibo

Destiny laboratory

Harmony (Node-2)

Entering European space

The next morning, 12 February, Peggy Whitson and Hans Schlegel worked to open the hatch to Columbus. The module was set to "Berthed Survival Mode" with power, and some computers were online, but not all systems were activated. With no ventilation or life-support astronauts were not allowed to stay in the lab for long periods of time.

The hatch was opened at 14:06 GMT and the two European astronauts, Hans and Léo, had the honour of being the first to float into the new module.

That Columbus was considered European was based on who had funded and built the laboratory. The Space Station partners agreed that each module was to be registered at the United Nations and 'local' laws would apply inside that module. Astronauts are bound by a code of conduct to simplify matters but in theory a different law applies in each module on the International Space Station; American modules follow US law, while the Japanese or Russian modules have their own law.

Back in Columbus, having made sure there was no smoke from any fires that could have occurred during launch, Hans and Léo got the ventilation system running inside the lab.

The duo wore masks and safety goggles as a further safety precaution, as the vibrations of launch could have caused small particles to circulate. On Earth a quick wipe with a damp cloth will remove most dust but in space it can irritate eyes and lungs, as it circulates indefinitely until filtered out by the ventilation system.

The Columbus survival mode setup was finished and the next step started immediately: final activation. Hans commented in an interview later during the mission that the module "smelt like a new car".

The 24-step method

Though the ventilation system was now running, the laboratory was still far from ready to use.

The Columbus control centre had a 24-step checklist to run through, starting with the cooling system and including running data cables and switching on computers and external heaters.

Now that Columbus had basic power and could be heated it was important to set up the cooling system so the computers inside would not overheat. Léo put on his space-plumber gloves and with Peggy's help the two astronauts connected the pipes that ran from Harmony to Columbus.

While everything was going smoothly above Earth, Munich Control Centre had a problem. The main data connection for vital Columbus systems was failing to start. As with most important systems in spaceflight, two models of the same system are included by design — if one fails the other can take over and this had happened automatically on the fresh module.

For ground control this meant that Columbus was already running on backup systems — not ideal for a freshly installed space laboratory. After some cautious deliberation, the activation was allowed to continue nevertheless.

Once the water circuit was connected another problem popped up immediately: one of the valves was leaking water. Gingerly, the teams on Earth decided to apply the age-old technique of turning it off and on again. They sent the command to close the valve and open it a few times in succession, hoping it would seat better and knock into place. The gentle nudge proved to work and the cooling system stopped leaking.

Step two was to turn on the computers that would regulate the cooling system. A catch-22 here was that the computers themselves needed the cooling system, so from the moment they were turned on they had four hours to get the water circulating before they would overheat.

By now mission control was getting used to problems and this step did not disappoint. Computer screens on Earth flashed with warnings: temperatures were far too high.

The problem turned out to be a classic case of overlooking an aspect of a routine operation some months before. The International Space Station had switched from one set of operating computers to another and emergency heaters in the modules were automatically turned on to supply heating during the brief switchover. As soon as the new computer booted up it turned all the heaters off again, but as Columbus was not installed yet the command to turn its heaters off never arrived. For months the command for the emergency heater was set to on, but there was no physical heater in place so nothing happened. Now that Columbus had been attached to the Station, the emergency heating kicked in.

The fix was as easy as turning off the heat exchanger and waiting for the water to cool down before starting the Columbus cooling system again.

At 21:37 GMT the pumps were switched on again. The control centres could relax... until the next problem.

Further problems

As the Space Station hummed in inactivity with dimmed lights, the next step on the checklist continued to give trouble to the mission control night shift. The mission management computer that would interact with the American part of the Space Station did not start up without errors and had to be rebooted before it would behave.

The flight controllers on Earth decided to plough on and activate another computer. Here too errors came in; Columbus was simply not accepting commands to its normal operating computers, with only the vital life-support computers functioning as they should.

After two reboots failed to solve the issue, a transatlantic "anomaly resolution team" came together.

NASA colleagues provided a clue: their main computer reported that its memory was so full it was no longer sending new commands. A solution could be to send a "clear memory" command — an act that required agreement by all international partners. Sending a wrong command risked having the main Space Station computer fail, a computer that controlled almost every aspect of the outpost.

A second, less dangerous, suggestion was to switch over from the operational computer to one of its many backups — though this would involve losing contact for longer than anyone on Earth liked.

Both options were being considered as the astronauts slept. Rex and Hans were once again in the airlock prior to heading outside for the mission's second spacewalk the next day. In the morning, mission control decided to return Columbus to a "safe berthed mode" with the exception of two ventilator fans. They informed the astronauts when they woke up that unfortunately the European laboratory could not be used as planned until the computer problem had been solved.

Remove after flight

On 13 February, most of the team set about installing laptop computers and experiment facilities in Columbus. These laptops allowed commands to be sent directly to Columbus's main computers, bypassing the American system that was returning the "memory full" errors.

A large part of getting Columbus ready was spent removing retainers that kept hardware in place. Due to the intense nature of a launch, numerous retaining screws had been used to keep sensitive equipment from being shaken about too much during the acceleration period. These screws serve the same purpose as when transporting a washing machine or dishwasher and must be removed before use. This required ripping up the floor of Columbus and locating the screws in a tangle of wires – a time-consuming task, but one that didn't require computational power.

Outside the Space Station, Hans and Rex replaced the empty nitrogen tank. While they were out, Hans also moved over to Columbus to remove the bolts that had kept it secured to the Space Shuttle. They were no longer needed and were acting as a heat-bridge, reducing the effectiveness of the module's insulation.

The spacewalk lasted over seven hours, during which time Léo and Dan Tani installed the Biolab experiment facility inside Columbus. Turning the facility on, however, would have to wait a while longer.

I consider the work we did during Expedition 16 to prepare for the arrival of Columbus one of the most significant accomplishments to date for a 3-person crew on the International Space Station. Moving Node-2 and conducting many spacewalks made room for Columbus and the Japanese modules. The culmination of our hard work with the arrival of Columbus on 9 February was one of my best birthday presents ever.

Peggy Whitson
NASA astronaut

The switch

On Earth, ground control continued to work on the computer problem. After running several risk assessments, they were confident that switching to one of the Station's backup computers was safe and would solve the memory issue. Not wanting to be cavalier, they waited until the spacewalkers were back inside (Hans and Rex closed the hatch at 21:12 GMT) before ground control started the switchover procedure around 22:00 GMT.

During the switchover, contact would be lost with the Space Station for a few tense moments as the backup computer started up. With a great amount of trepidation, Columbus control centre sent a test command to its module... the laboratory's main computers were finally online, and installation could resume in earnest.

A week in space

After seven days in space working non-stop to connect Columbus, the Shuttle and Space Station astronauts were allowed some rest. The planners programmed a light day on 14 February; in the evening, the astronauts all gathered in Columbus for its first interplanetary press conference. German Chancellor Angela Merkel was one of the guests on Earth. The two-way live transmission from the space laboratory to the press event in Germany showed that Columbus's video systems were running as planned. Live television is now so common we hardly think about it, but the infrastructure to send video and sound from Columbus to Earth and back was set up in seven days, a testament to the network engineers and satellites that work for the Space Station.

At night Rex and Stan slept in the airlock again. The next day, they were heading outside for the third and final spacewalk to install two facilities on Columbus.

The EuTEF

The European Technology Exposure Facility – or EuTEF for short – was the first exterior facility installed on this last spacewalk. Instruments measured radiation, atomic oxygen, micrometeorites, electrostatic forces and the temperature to get a better idea of the environment the International Space Station was flying in, and help design better protections for spacecraft in the future.

A second element of EuTEF contained some scientifically approved stowaways. Researchers had chosen a number of living organisms to travel around Earth on this third-class ticket. Lichens, fungi and other "extremophiles" that were considered hardy enough to survive space travel unprotected had an epic journey ahead of them.

Whereas Rex and Stan worked in the relative luxury of a spacesuit, the vacuum of space was sucking out the water, oxygen and other gases in these organisms. Their surrounding temperature dropped to −12°C as the Station passed through Earth's shadow and rose to 40°C in sunlight. In addition, these extremophiles were getting the full blast of the Sun's radiation without the layer of atmosphere that protects life on Earth.

Some 9,000 orbits and 18 months after Rex and Stan installed the facility, two astronauts on another spacewalk took EuTEF back inside and shipped it back to Earth on Space Shuttle Discovery. Researchers discovered that the some lichens had, in fact, survived the experience and were nurtured back to life from space hibernation.

ESA has a long history of testing organisms and organic chemicals in space. Follow-on studies with other samples placed outside the International Space Station showed that "water bears" and organisms used to brew kombucha tea also survive spaceflight unprotected. These findings hint at the possibility of species colonising planets via meteoroids, a theory called panspermia. Could life on Earth have started when organisms on an asteroid hit our planet to kick-start evolution as we know it? No one knows for sure but the exposed facility that Rex and Stan installed has proven that this is indeed a possibility.

We had tested all the infrastructure, ran lots of simulations with the EuTEF Ground Model and we were feeling totally confident and prepared. Nothing would excite or surprise us – or so we thought. But operating the actual EuTEF in space(!) was far more exciting then we could have ever expected. Just imagine: you are sending a command – and you get the actual science data, not just a simulation. Getting the first set of images for example: they were overexposed and had a strange colour tint but they were the real ones and the first.

Volker Koehne
Payload manager

Installing the Sun monitor

The second box installed on the spacewalk consisted of three instruments to monitor the Sun's energy output. Earth-bound observations of sunlight are hindered by our atmosphere, which filters sunlight by wavelength – one of the reasons we see the sky as blue. To get a clear reading of the energy our closest star emits, researchers wanted to monitor it from beyond Earth.

Once properly set up, the 75-kilogramme box perfectly tracked our Sun continuously for almost a decade before it was switched off in February 2017. During its operating life, it recorded 12,000 sunrises and sunsets. The team behind it even managed to convince the Space Station operators to rotate the entire outpost. In 2012, the 450-tonne Space Station, the size of a small block of flats, was rotated all to aid the refrigerator-sized solar facility attached to it.

Solar is not sexy science; the monitor delivered no pictures, but ran silently collecting numbers in the background. These numbers were used amongst other things to improve computer models of our Sun, helping advance climate predictions and understand the impact of Earth's main source of energy.

When Rex and Stan installed Solar, the instruments were set to work for 18 months. It was switched off nine years later having delivered its goals: improving the accuracy of solar spectrum reference data down to 1.26%.

So long Atlantis…

For the crew of STS-122 the mission was coming to an end. On 16 February, the full complement of humanity in space – all 10 astronauts, comprising four different nationalities – finished the installation of Columbus.

While the crew worked on the final elements of the installation, Space Shuttle Atlantis fired its engines for 36 minutes to boost the Space Station's orbit. Although in theory Newton's cannonball would orbit Earth indefinitely once it had enough speed, the International Space Station flies at a relatively low altitude of around 400 kilometres. For comparison, the satellites that power your navigation system are placed 22,000 kilometres above our planet, and communication satellites even further – orbiting our planet roughly 90 times further away than the Space Station.

At 400 kilometres above our planet there are still traces of our atmosphere that gradually slow down the Space Station and any other object orbiting Earth at that height. Over time the speed drops slightly and the orbit too, making timely reboosts a necessity to avoid falling back to Earth. The Space Station's orbital height is a trade-off – the lower it flies the easier it is to get to but more boosts are needed to stop it from falling to Earth.

When the time comes to retire the International Space Station it will be carefully deorbited to disintegrate over the oceans where there is no chance of debris hitting people. All satellites in a low-Earth orbit eventually fall back to us, most disintegrating harmlessly in the atmosphere. Solar radiation increases atmospheric drag, so when the Sun is more active more satellites fall back to Earth. The Solar facility that Rex and Stan installed has improved predictions of when these older satellites will tumble into our atmosphere – and eventually this data will be used for when the International Space Station is laid to rest.

Final farewells

On Sunday 17 February, after emotional farewells between the departing crew and the three astronauts who would be staying on in space, the hatches closed between Atlantis and the International Space Station. Dan Tani said farewell to his pied-à-terre in space and left his spot for Léo, who remained to continue working on Columbus.

Overnight the Space Shuttle transferred over 40 kilogrammes of oxygen to the Space Station. A further 635 litres of water had already been reassigned as basic supplies for the remaining astronauts.

Atlantis undocked at 09:24 GMT. The Shuttle pilot, Alan Poindexter, saluted the Columbus module with a final farewell, performing a full fly-by of the Space Station to get a view of the new European contribution. The outpost had grown by almost 30 m^3; the mission had lasted 12 days, 18 hours, 21 minutes and 40 seconds.

COLUMBUS MODULE

FRONT VIEW

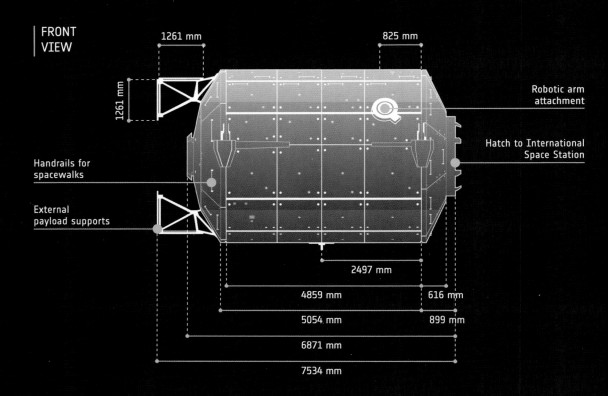

1261 mm

825 mm

1261 mm

Robotic arm attachment

Handrails for spacewalks

External payload supports

Hatch to International Space Station

2497 mm

4859 mm

616 mm

5054 mm

899 mm

6871 mm

7534 mm

TOP VIEW

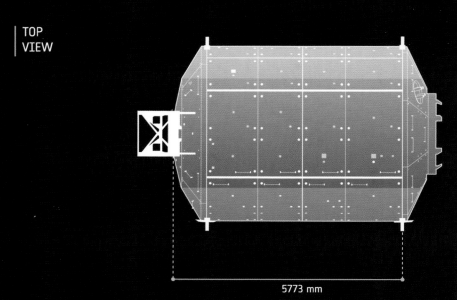

5773 mm

STARBOARD
VIEW

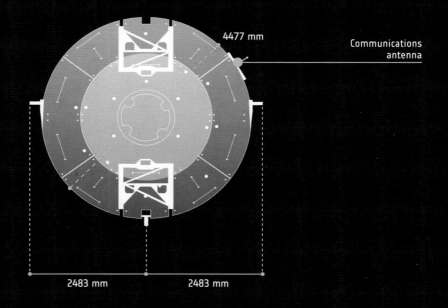

4477 mm

Communications
antenna

2483 mm   2483 mm

R 2623 mm

PORT
VIEW

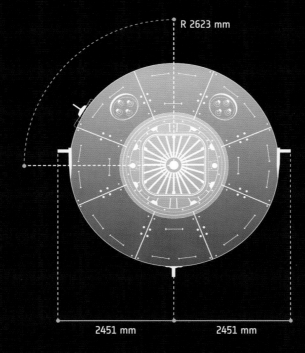

2451 mm   2451 mm

HOW IT WORKS

Columbus is a part of the International Space Station and shares its life-support systems. These include air to breathe, electrical power, ventilation and temperature control – the basic but vital elements needed to keep hardware running and, of course, astronauts alive.

Power!

Columbus gets its power via electrical wires from Node-2. All power comes from the Space Station's enormous solar arrays. The solar cells alone cover almost half the area of a football pitch (about 2500 m^2) and each solar "wing" is longer than a jumbo jet's. In direct sunlight the system easily provides enough electricity to power 40 households.

The Space Station's solar array has some advantages over Earth-bound solar panels. With no clouds in space, it receives unfiltered sunlight, plus every 45 minutes, the outpost passes from sunlight to darkness and back again. The shorter day–night cycle means the Space Station's batteries do not need to store electricity for a 12-hour night as on Earth.

The downside is that the solar panels need to track the Sun perfectly as the Space Station speeds through its 45-minute window of sunlight. The arrays are on gimbals that track our star to get the most of their time in the sun. Like giant sunflowers, once the Space Station passes into the shadow of Earth the solar arrays turn back to prepare for the next salutation. The system has powered the Space Station since long before Columbus arrived and has now tracked the Sun like this over 50,000 times, a testament to its engineering and build quality.

A second disadvantage is that the large surface area of the solar array acts as a net for space debris. Travelling at 7.6 km/s through space, even a small fleck of paint can cause serious damage to satellites and the Space Station. The Cupola bay window on the Space Station, for instance, has cracks in it from where dust particles have hit during use. The window is quadruple-paned by design and shutters are closed when it is not in use as an extra precaution.

Space agencies track space debris and small particles continuously from Earth, so if something is on a collision course the Space Station can fire its thrusters to move out of the way. In serious cases, astronauts shelter in their spacecraft for safety, and if needed they can abandon ship and return home.

Dust can impede the effectiveness of solar panels on Earth but a quick clean will get them operating at full capacity again. Spacecraft solar panels dealing with dust projectiles become punctured with holes that inevitably lead to reduced output. The large solar arrays also provide more drag – slowing down the Space Station ever so slightly, making periodical orbital boosts a frequent necessity.

The Space Station's electrical system supplies a reliable current to all modules and systems through 12 kilometres of cabling. There is a maximum amount of electricity that can be generated, of course, and it does vary, so how the power is divided between experiments is frequently a subject of heated discussion. Each experiment and space agency needs to consider the amount of electricity it can consume and every module has a list of priority equipment.

When power is reduced for some reason, the Station's computer consults the "load shed table" to decide what equipment is switched off one by one until there is enough power again. First come non-essential items such as public webcams and external heaters that can survive for hours without power, then the scientific experiments. The list is highly political and no researcher wants their experiment anywhere near the top.

Temperature

Temperature is controlled by electrical heaters and ammonia-filled radiators that look like wavy versions of the solar arrays, but come in a heat-exchanging white colour (for the same reason your refrigerator is probably white too: it reflects heat well).

The Space Station's system is not unlike the cooling system in a car. A cool liquid circulates past the hot engine, warms up and then passes to a radiator where it cools down again. The cooler liquid is pumped back to the hot area, and the process repeats, keeping the engine from overheating. In cars, water is used as coolant, but people who live in colder climates put anti-freeze in the system to avoid the water freezing and damaging the engine. The Space Station uses ammonia as its coolant because it has a very low freezing point of −77.7°C. Liquid ammonia is its own anti-freeze and better suited to circulate through outside radiators in the cold conditions of space.

The one big disadvantage of using ammonia is its toxicity to humans. Exposure to high

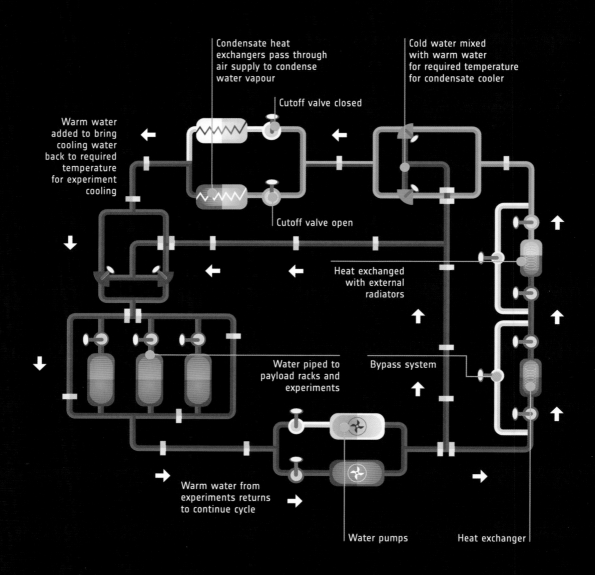

Condensate heat exchangers pass through air supply to condense water vapour

Cold water mixed with warm water for required temperature for condensate cooler

Cutoff valve closed

Warm water added to bring cooling water back to required temperature for experiment cooling

Cutoff valve open

Heat exchanged with external radiators

Water piped to payload racks and experiments

Bypass system

Warm water from experiments returns to continue cycle

Water pumps

Heat exchanger

concentrations of ammonia can cause freeze-dry burns, tissue and eye destruction, convulsive coughing and, unsurprisingly, death. Great care is taken to avoid ammonia leaking into the Space Station, so the system only circulates ammonia outside and passes its cooling properties to an internal, water-based, cooling system for the computers and experiments.

The water that flows through Columbus exchanges heat with the ammonia circuit in the Harmony module. As cooling is so important, most of the system is built in duplicate — and some elements even in triplicate. If ever a pump or heat exchanger fails, the system continues working while an astronaut can inspect and repair the faulty equipment.

As water passes by the ammonia heat exchanger it is cooled to between 2.1°C and 4.2°C. This temperature is precisely set to the level where water vapour in the air condenses, so the system not only cools the air in Columbus but also captures water droplets exhaled by astronauts for reuse as drinking water.

The atmosphere in Columbus, as in the rest of the Space Station, is kept at a cosy 18°C to 20°C, which is why you regularly see astronauts in T-shirts in space. The temperature has been known to increase, however, at the request of some astronauts who are more accustomed to warmer climates on Earth.

Air conditioning

The air in Columbus is shared with the other modules that make up the ISS and circulated continuously in a closed system. More than just a luxury, the air conditioning system provides important life-support functions.

Gravity is integral to all sorts of phenomena that we take for granted on Earth. The air in our homes circulates naturally: hot air rises because it is lighter than cool air — the basic principle behind hot-air balloons. In weightlessness, nothing is lighter than anything else, so earth-like convection does not occur. Without ventilation, a sleeping astronaut could suffocate in the carbon dioxide that pools around their face.

As air from Harmony is sucked into Columbus it passes through a filter to help clean the atmosphere. As air is recycled indefinitely on the ISS, many astronauts remark on the pungent smell when they first arrive: imagine how a passenger jet would smell that has not landed for 15 years and you might get the idea. Microbes, dust, chemicals and yes, man-made odours, circulate continuously until they are very slowly removed by the filter system. Before anything is sent to the Space Station extensive tests are done to make sure that the material will not pollute the artificial atmosphere. On Earth, the smell of cheap plastic from a freshly unpacked toy disappears quickly as it evaporates into the room, but on the Space Station the plastic chemicals accumulate and can cause health hazards.

The air filters have a happy extra function: retrieving lost equipment. Astronauts on the Space Station are constantly losing things, not because they are careless (quite the opposite) but because all objects unattached will gently float away. Astronauts report finding objects weeks after losing them near ventilation fans – from ear plugs to pills, Christmas decorations and more. Keeping track of the items in space is so demanding that special teams in control centres are formed to keep an eye on inventory.

The European space lab's ventilation system can be shut off from the rest of the Space Station in an emergency. With the hatch to Harmony closed, air inside Columbus can continue to circulate to locate a leak or contain a fire.

Four valves around Columbus can open to let air escape into space in a worst-case scenario, freeing the module of toxic chemicals or extinguishing any fire. As there would be no air left, though, the hatch would need to be closed first or everything in the Space Station would be sucked out – including the astronauts.

The Dutch company behind the fans in Columbus takes some pride in having produced the quietest fans on the Space Station. Many astronauts have remarked that Columbus is the most relaxing place on the outpost – a quiet escape from the incessant hum that comes from living in a large spacecraft.

Life-providing oxygen is created in other places in the orbital outpost. Fresh supplies of oxygen are brought up by cargo ships and there is a backup system that produces oxygen through chemical reactions. A third way to get oxygen is to separate the oxygen and hydrogen in water with electricity. The hydrogen is sent into space or recycled. Carbon dioxide is scrubbed from the air by the Russian Elektron machine and shot out to space or recycled to make water.

With everything to keep the astronauts alive and the experiments powered all that is left to have a working space laboratory is the experiments and their processing facilities.

COLUMBUS VENTILATION SYSTEM

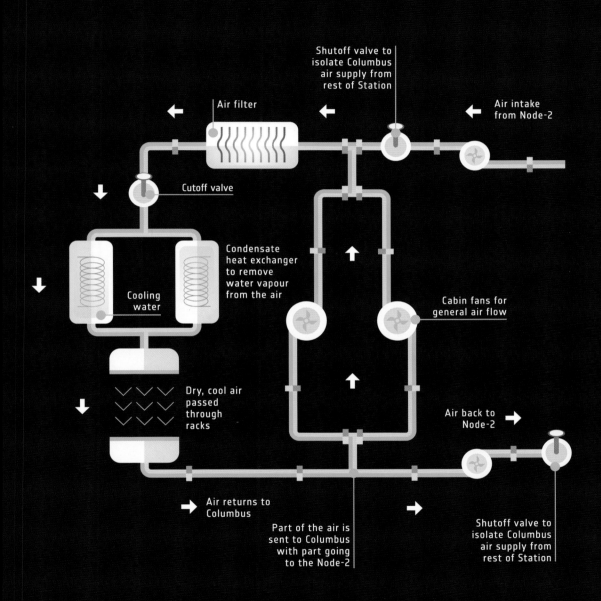

Shutoff valve to isolate Columbus air supply from rest of Station

Air filter

Air intake from Node-2

Cutoff valve

Condensate heat exchanger to remove water vapour from the air

Cooling water

Cabin fans for general air flow

Dry, cool air passed through racks

Air back to Node-2

Air returns to Columbus

Part of the air is sent to Columbus with part going to the Node-2

Shutoff valve to isolate Columbus air supply from rest of Station

ATV | AUTOMATED TRANSFER VEHICLE

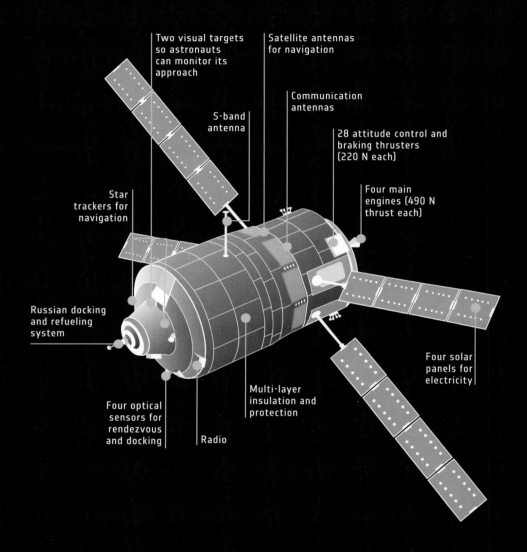

Two visual targets so astronauts can monitor its approach

Satellite antennas for navigation

Communication antennas

S-band antenna

28 attitude control and braking thrusters (220 N each)

Star trackers for navigation

Four main engines (490 N thrust each)

Russian docking and refueling system

Four solar panels for electricity

Four optical sensors for rendezvous and docking

Radio

Multi-layer insulation and protection

I spent 200 days on the International Space Station and in that time, with my crewmates, I tried to take good care of humanity's outpost in space. However, as a European astronaut, I felt a special responsibility and affection towards Columbus, a piece of European real estate in orbit.

Samantha Cristoforetti
ESA astronaut

Transport

The European Transport Carrier is a storage facility and work bench in Columbus that enjoyed a previous life as a transport container in Space Shuttles before finding a home in Columbus. A standard-sized rack (it can carry up to 410 kilograms of equipment), it allowed for regular supply and return of all experiment and service items.

Together with the European Drawer Rack these storage spaces are important to all the experiments that take place inside Columbus; they support Biolab, the Fluid Science Laboratory and the European Physiology Module by offering storage space for astronauts.

Keeping the International Space Station stocked requires frequent supply trips, and Europe's contribution was the Automated Transport Vehicles. Five spacecraft flew at 18-month intervals from March 2008 to July 2014, stocked with cargo, experiments, food, oxygen, fuel and water. The uncrewed spacecraft were launched from French Guiana and docked with the Russian section of the Space Station automatically. As well as delivering supplies, they also had the power to boost the Station into a higher orbit and move the outpost out of the way of dangerous space debris. At the end of their mission, astronauts would load the spacecraft with waste that would burn up on re-entry in Earth's atmosphere – a high-speed high-altitude space incinerator.

EDR | EUROPEAN DRAWER RACK

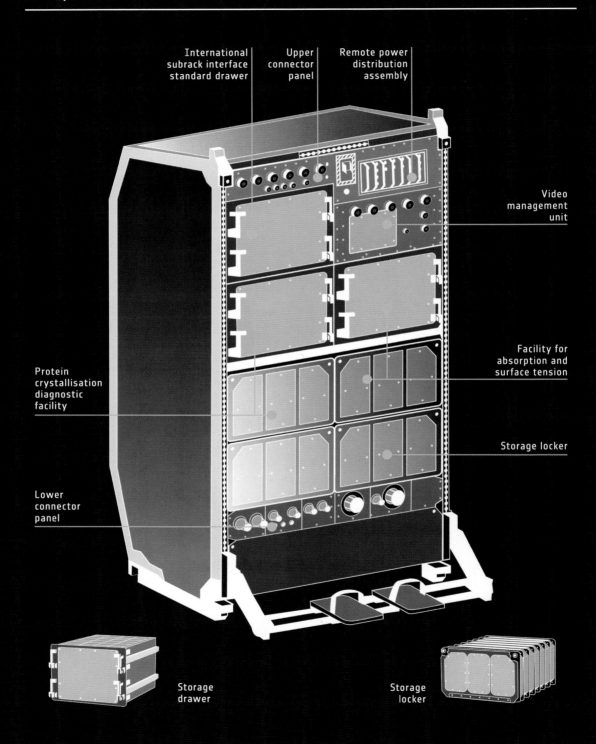

International subrack interface standard drawer

Upper connector panel

Remote power distribution assembly

Video management unit

Protein crystallisation diagnostic facility

Facility for absorption and surface tension

Storage locker

Lower connector panel

Storage drawer

Storage locker

A DAY IN THE LIFE

The Columbus laboratory is an integral part of the International Space Station and can be used by all the astronauts on board. With this said, it is rarely ever packed full. Generally the schedulers on Earth will give astronauts as much working room as possible and plan to have them work separately in each module — astronauts have remarked that the Space Station is such a large place they will occasionally go hours without seeing their floating colleagues.

A typical day in space hardly exists but this overview aims to give an idea of the tasks run in Europe's real estate in orbit.

Morning

Astronauts have telephone-booth-sized quarters with a sleeping bag and room to store personal items — including a laptop. The laptop is connected to a computer on Earth that tunnels internet securely so they can check emails and keep in touch with those back on Earth. The connection speed is nothing to write home about — like the pre-broadband days, websites typically take over five minutes to load.

Morning rituals are mostly the same as on Earth but great care is taken to not let shaving cream and hair float away. As there is no sink or running water, wet wipes make do as a space shower. Astronauts often list showers and fresh air as things they miss most in space. Oral hygiene is important wherever you live so astronauts brush their teeth twice a day as usual — except they swallow the toothpaste instead of spitting.

Each day on the Space Station starts with a mandatory meeting for all astronauts. The 20-minute "Daily Planning Conference" goes over the day's planned activities and any changes to the schedule from mission control. The daily schedules look like the programme of a large music festival, with each astronaut's tasks displayed in horizontal blocks; the longer the block the more time a task will take. The overview also includes the brief periods when the Space Station loses contact with ground control and special requirements such as closing the shutters of the Cupola to protect the windows. Each task and item can be double-clicked for more information and includes detailed step-by-step procedures with notes, warnings and requests. Astronauts can add their own comments for the teams on Earth to improve operations for next time or just say "Thank you".

Many experiments ask for bodily samples, from blood and urine to faeces and saliva, and many are taken before breakfast for the same reason hospitals and doctors often require you give a blood sample on an empty stomach. Taking a blood sample in space requires no extra equipment than used on Earth and astronauts are trained to draw their own blood. Keeping the needle, tubes and plasters from floating away requires some extra coordination and use of Velcro on the walls. Blood samples often go in a centrifuge such as in the Biolab facility in Columbus to separate their parts before cold storage in the European-built −80°C freezers that are located in the Japanese and US laboratories.

Breakfast can be anything that doesn't crumble and cause a mess or can be contained in a bag or tins. Fresh milk is not an option. Coffee and tea comes powdered in bags with a straw that are filled with warm water from a dispenser.

Going to the toilet is much the same as on Earth — except there are separate suction cups for numbers one and two. Like stray hairs from space haircuts, the suction keeps the ablutions from floating away.

On a weekly basis ESA astronauts in space have conferences with the Columbus Control Centre and can also consult with their private flight surgeon to discuss how they feel and warn of any medical issues. European astronauts run these meetings with Earth from the Columbus lab. As on Earth, doctor–patient confidentiality is strictly enforced and control centres do not listen in. As astronauts' immune systems are weaker in space, great care must be taken to avoid any lingering problems. Due to fluids collecting around the head, the regulatory glands there assume there is too much blood in general and send signals to reduce blood production across the entire body — an astronaut loses over a litre of blood while in space. People start to faint when they lose 2 litres of blood so astronauts have very little margin. Even small cuts are a serious health hazard.

As a European, I felt that Columbus was my "special place" on the International Space Station. I was proud when we performed experiments in our orbital lab, which happened quite often, as most of the human physiology research is performed there. Initially I was surprised that the module always seemed in disarray – but then I realised that's the sign of a place that's being constantly used, and it never bothered me again.

Luca Parmitano
ESA astronaut

Before lunch an educational radio contact with a school somewhere around the world brings astronauts closer to schoolchildren. The Amateur Radio on the International Space Station organises frequent talks over simple radio waves as the orbital outpost flies overhead. Schoolchildren have up to 15 minutes to ask their questions before the Space Station flies out of range.

Setting up the equipment is easy and requires only a few minutes. The antenna used for these calls is on the outside of Columbus and all astronauts are trained to use the ham-radio. Enthusiasts can try hailing the Space Station themselves with regular equipment – with a bit of luck an astronaut will reply to your call.

Midday maintenance

Lunch is generally eaten quickly to get back to work or spend some free time looking out of the Cupola windows. There is no fixed time for lunch, so on most days an astronaut will grab a bite to eat on their own whenever they feel hungry.

In the afternoon, astronauts typically get on with further experiments. For instance, they could be assigned to swap an experiment container in the Electromagnetic Levitator with alloys such as titanium and nickel and check that the connections are working. This generally takes an hour and once done the control centres on earth can take over issuing commands to heat and cool the metals inside.

As the experiment continues unattended, the astronaut might maintain Biolab by changing the centrifuge belt to prepare for an experiment that will arrive on a supply spacecraft shortly. Biolab is a complicated and large machine but without weight the unit can easily be pulled out of its rack for inspection and maintenance.

Another astronaut could be collecting the portable incubator centrifuge Kubik from the European Drawer Rack to power it up and insert cell tissue for a genetic study. Cells are often spun in the centrifuge at different gravity levels and frozen chemically at different times to see how they react in space. Setup and inserting new tubes with cell cultures takes around an hour of time. Depending on the type of experiment, the incubator might also need some time to warm up.

A third astronaut might return a 3D-printed Columbus toolkit to its storage location after borrowing a tool. Clips keep the tools from floating away and future long-duration missions could carry their own 3D printers in space to print out missing parts immediately.

With life-support equipment running all day and night it should be no surprise that maintenance is an important part of daily life in space. On any day, an astronaut can be asked to check the condensate exchange system. First of all, a sample is taken of the water that collects for recycling. Not unlike a student dormitory, six people living in a closed system with recycled air and water inevitably offers a nice habitat for mould and bacteria to grow. These are monitored for obvious reasons, with regular samples taken and checked for toxicity. Disassembly, inspection, replacement of a part and starting the system up again takes around two hours. Astronauts can leave "flight notes" for mission control on each task — these observations can help keep track of equipment or improve scheduling for next time and are used as digital sticky notes.

Testing drinking water is a much quicker task thanks to a multi-layered petri dish developed in France with a powdered growth medium. When water is added, microbes form coloured spots revealing their locations. A picture taken with a smartphone or tablet is processed immediately to calculate how contaminated the water is — with no other equipment necessary, the task takes only 25 minutes per two samples. Developed for the International Space Station, the petri dish is clearly useful on Earth, especially in disaster areas, where water could be contaminated — a picture and calculation is much cheaper and faster than sending samples to a laboratory.

Evening

After a long afternoon session running human physiology research in space, a special evening on the International Space Station will be getting the go-ahead to pick a fresh crop of lettuce from NASA's greenhouse. Half the harvest is stored for analysis back on Earth but the other half can be eaten as a rare fresh accompaniment to space tacos. It takes an hour to collect the tools, harvest the crop, pop half in a bag and then into a freezer — and clean up the greenhouse for the next batch.

The evening is free for astronauts and blocked out as "pre-sleep" in the schedule. This period allows time for astronauts to look at the schedule for the next day, to brush their teeth, look out of the window or phone home before zipping into their sleeping bags.

This overview leaves out the many hours spent on emergency drills, unpacking cargo from newly arrived cargo vessels, refresher courses, or the mandatory 90 minutes of exercise every astronaut does each day on the Station's four exercise machines to stay fit in space. When a spacecraft arrives or a spacewalk is planned, the normal schedule is discarded as all hands prepare tools, spacesuits and procedures for the big event.

Like living and working in your own home, the daily chores such as cleaning dishes, vacuuming, repairs and maintenance can take up a large part of the day – astronauts outsource as much as possible to ground control but there is only so much you can do when you are a minimum of 400 kilometres away and not on the same planet. All unpacking of supplies, repairs and cleaning end up being in the hands of the astronauts.

Saturday is cleaning day on the International Space Station (unless a supply vessel is arriving) and everything gets a good wipe. One chore astronauts are spared is laundry: with no running water or washing machine, clothes are thrown away when dirty.

At night the Space Station switches off its lights leaving Columbus to glow in the emergency lighting. As the astronauts sleep in their crew quarters, experiments in Columbus continue incubating, bubbling, melting or spinning under the watchful eyes of the night shift at ground control.

It is a unique feeling to be inside Columbus, but I might have a bias towards that module. It is a neat facility and it was such an honour to be part of bringing it to the Space Station. One of my favourite memories of working in Columbus was when I returned to the Station on the STS-135 mission. Ron Garan and Mike Fossum were scheduled for a spacewalk and I grabbed them to do the revision for it in Columbus.

Rex Walheim
NASA astronaut

PHOTO GALLERY

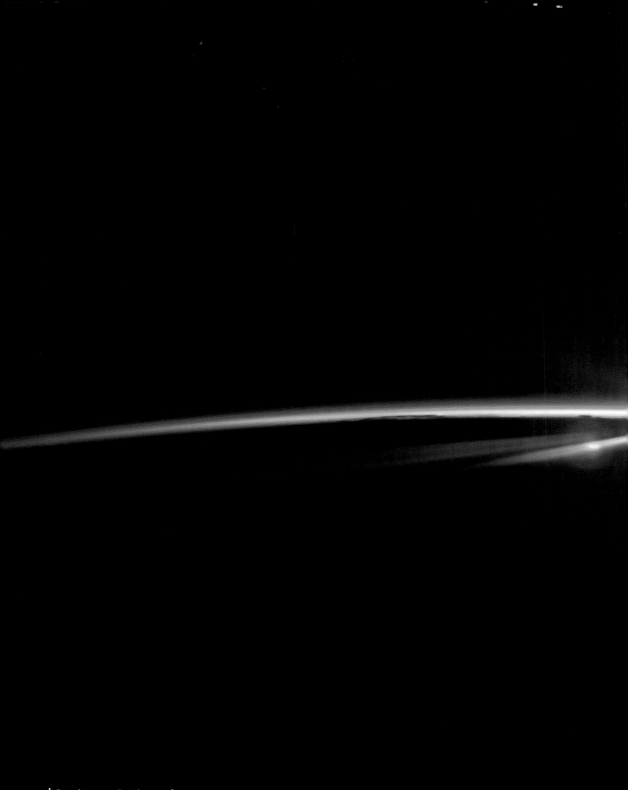

Sunrise over Earth seen from
International Space Station

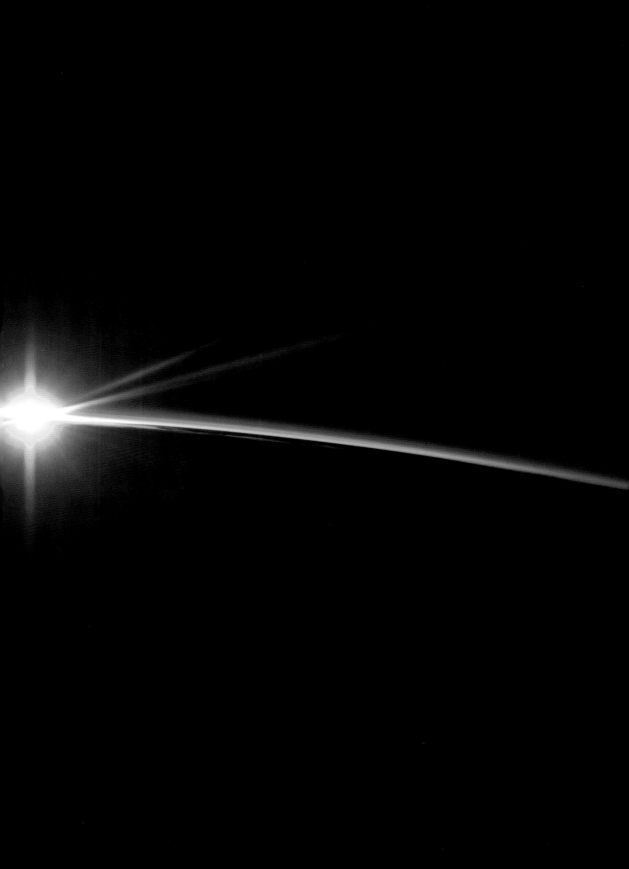

Installing Columbus
with the robotic arm

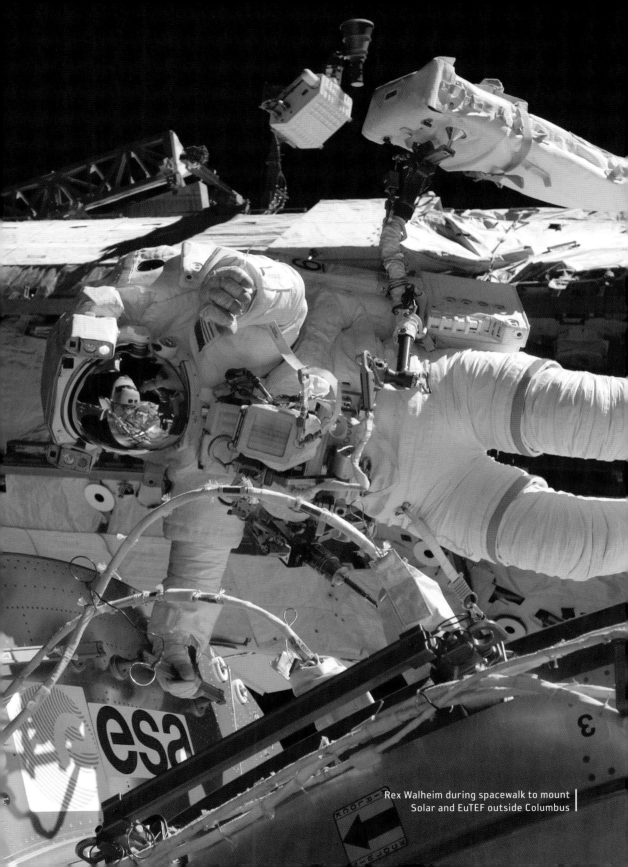

Rex Walheim during spacewalk to mount
Solar and EuTEF outside Columbus

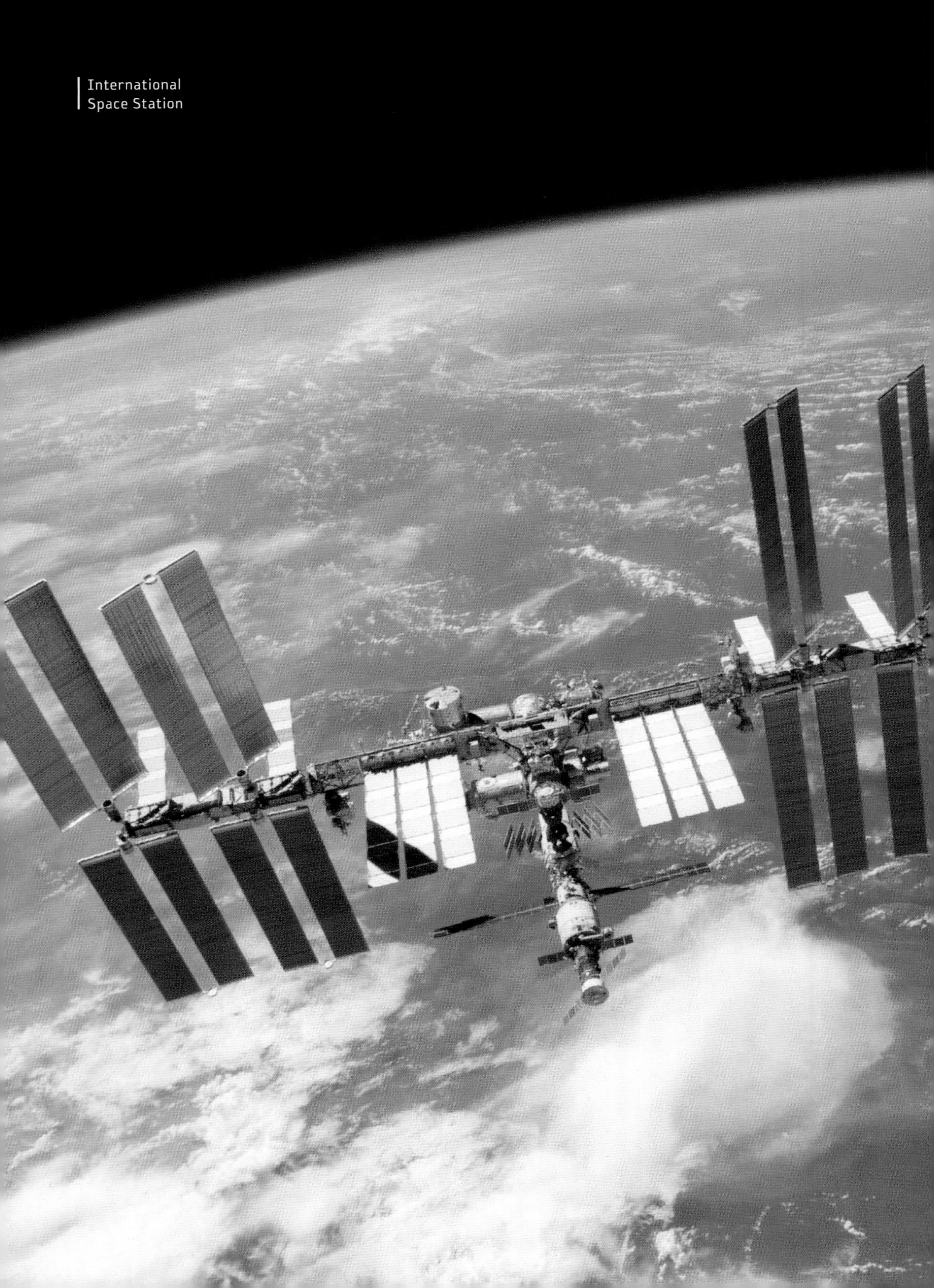

International
Space Station

Full crew of Space Shuttle and Space Station during Columbus installation mission

Spacewalk

International Space Station with Automated Transfer Vehicle (right) and Space Shuttle

Rex Walheim during
spacewalk

Columbus

ESA astronaut Samantha Cristoforetti
working on Biolab

SCIENCE

What makes humans unique over other animals is their harnessing of technology to travel and explore new places. Deep-sea divers or mountaineers would not go far without technology to survive underwater or on cold and icy mountain tops.

As we wish to explore further — and we are a planet of natural explorers — we need to develop the know-how and technology to survive and thrive in space.

When Edmund Hillary first reached the summit of Mount Everest he was aided by then-new technology such as rubber oxygen masks. These technologies were designed and developed in collaboration with the alpinists that were pushing human exploration to new heights at the time, but the only way to make sure they worked was to put them to the test in the field — or even better, the highest mountain top. Human spaceflight is no different, just 400 kilometres above sea level instead of 8 kilometres on top of Mount Everest.

In addition to understanding how the human body reacts to space we also want to see how materials and equipment fare in space. The Columbus laboratory offers European researchers and engineers a place to put their design to the ultimate test — if it works in space it will work anywhere.

All the work to get Columbus up and orbiting our planet safely would be pointless without the sophisticated facilities inside.

Columbus accommodates the international space standard experiment rack size in all modules so experiments can be easily installed and moved. Columbus is the smallest of the space laboratories but it still manages to offer the same payload volume as well as power and data handling as the other orbiting labs.

There are 16 racks organised in four sets of four that cover many scientific disciplines as well as offering storage space and place for one-off experiments. Every rack gets power, cooling and a connection to the Space Station network so experiment data can be sent to researchers on Earth.

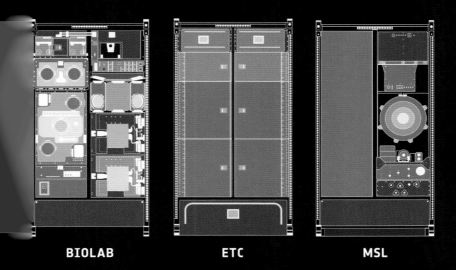

BIOLAB   ETC   MSL

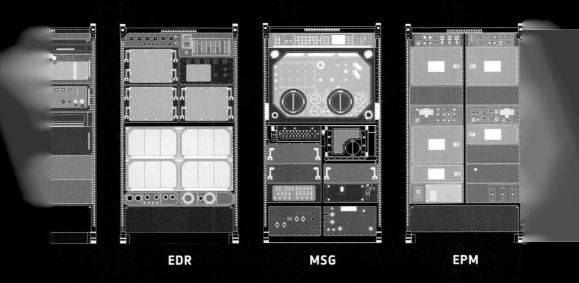

EDR   MSG   EPM

EDR = European Drawer Rack   EPM = European Physiology Module
MSL = Materials Science Laboratory   MSG = Materials Science Glovebox
FSL = Fluid Science Laboratory   ETC = European Transport Carrier

Science in space

European research in space covers many disciplines – thanks to the well-equipped Columbus laboratory and an inclusive selection policy for its experiments – but why conduct research in space at all?

A large part of the allure for scientists to run an experiment on Columbus is due to its lack of weight.

Any researcher can propose to fly their experiment on the International Space Station, but the European Space Agency considers each request based on how realistic it is and whether the proposals could be run elsewhere and still deliver useful results. The European Space Agency offers other platforms for research that allow experiments with gravity – from hypergravity centrifuges to 15-minute weightless periods in rockets that no human would survive, as well as safe-for-humans parabolic flights in refitted passenger aircraft that offer brief 20-second bouts of weightlessness. But Columbus is the only gravity research platform on offer that allows experiments to run for longer than 15 minutes at a time.

Why do we run experiments in weightlessness? To investigate how a laser works, you wouldn't run the experiment in bright sunlight but in a darkened room to observe the light without distraction. Researchers are always turning off external variables to focus on one aspect of their study. Whereas light, sound, vibrations, movement and so on can be "switched off" on Earth, the International Space Station is the only laboratory where you can "switch off" gravity.

The European Space Agency only runs research in space if it cannot be run anywhere on Earth. Broadly these experiments fit into two categories: experiments that aim to test reactions to being in space and experiments where gravity skews readings and hampers understanding of the finer elements of a phenomen.

Looking at humans

The first human to fly into space was Yuri Gagarin in 1961. Until he had orbited Earth for 90 minutes nobody could be entirely sure a human could survive a trip beyond our atmosphere. Aside from the force of being strapped to the top of 274 tonnes of rocket fuel and igniting it, astronauts have to live, breathe and eat without weight. Yuri survived and returned to Earth in good health, and ever since researchers have been learning about the human body and how it adapts to an environment that we have not evolved in.

In space everything from our biological clock to the fluid in our spines and inner ear loses track of its normal functioning. Bones and muscles waste away from lack of use as astronauts float through the International Space Station with a simple flick of the wrist – the legs used on a daily basis to walk are hardly needed in space. With less blood to pump and less physical exercise, even the heart starts to atrophy in weightlessness.

Whereas Yuri's first orbit of Earth was also his last – he returned to Earth less than two hours after launch – astronauts have now spent over a year in space in one go – long enough to fly to Mars and back.

As we gradually understand more about how humans react to spaceflight, we develop new ways to make astronauts more comfortable, from garments that stay fresh longer or offer electrically-induced exercise to trousers that use negative pressure to pull blood to the feet and recreate gravity's effect on Earth.

European research in this domain includes taking brain scans of astronauts to measure their "plasticity" – how quickly their brains adapt to new inputs. The reassuring conclusion is that brains adapt surprisingly well, though research is hinting that the effects of floating in weightlessness mark the brain forever. Hans Schlegel remarked that floating in weightlessness is like learning to ride a bike – once learnt you never forget.

Another experiment looked at astronauts' bones that go through accelerated osteoporosis during spaceflight. This disease costs Europe around €25 billion a year and typically affects the elderly, resulting in brittle and fragile bones with broken hips and arms from falls.

Studying astronauts in space showed that acidity in the body accelerates bone loss. Counteracting the acidity by eating less salt or taking simple bicarbonate pills provides a simple preventive measure. Whether this remedy works on Earth will require following large groups of people over their lifetime while monitoring their salt intake. Conclusive results will take many decades but space research is giving clues where to look for answers and recommending a healthier, low-sodium diet to avoid developing osteoporosis.

Not quite weightlessness

Earth is like all objects and attracts everything around it – the heavier an object the more it will attract. Experiments using extremely sensitive measurements have shown that a bowling ball placed near a ping-pong ball will exert force, pulling the smaller, lighter ball towards the heavier, but also that the opposite is true. On a minute scale the ping-pong ball attracts the bowling ball too, just significantly less. This works for all objects, including humans. If you jump in the air you will fall back to Earth due to gravity, but on a microscopic scale Earth also moves towards you.

This means that the International Space Station is not free of gravity. In fact, without Earth's gravity pulling on it, the outpost would career off into space never to be seen again. The weightlessness experienced by astronauts and their experiments occurs because all objects are falling at the same speed. You can recreate this effect by throwing a half-full bottle of water from one hand to the other in an arc. As the bottle travels upwards the water will stay at the bottom of the bottle, experiencing accelerated g-forces like astronauts during a rocket launch. Near the top of the bottle's curve the water will start to slosh about inside – all the forces on the bottle and its contents are equal at this point, it has enough upward momentum to counteract the pull of gravity and the water is suspended inside the bottle until you catch the bottle or it comes to rest on the floor. Gravity is now pulling the water against something – the bottle and your hand – and so weight is restored.

If you have trouble imagining this, consider being stuck with a bathroom scale in a lift that is falling from the top floor of a skyscraper. In your last seconds alive before being crushed at the bottom of the lift shaft you will be weightless. If you had the clarity of mind to perform a little experiment in your last moments, and tried to stand on the bathroom scale, it would be useless as both you and the scale would be weightless, floating inside the lift.

A more common experience, and one you will live to recount, is exploited at fairgrounds on rollercoasters. You know that feeling when your insides seem to float and your hair goes up on end? That is because for a few seconds you are weightless, your body and the rollercoaster are all in free-fall, just like the water in a bottle falling through air and astronauts in the International Space Station.

For this reason, scientists don't describe the Space Station as a zero-gravity environment, but use the term microgravity. Small changes in forces throughout the outpost can cause accelerations and g-forces in experiments. These perturbations can come during reboosts from supply spacecraft, minute adjustments to the Station's position, reverberations running down the hull and even the footsteps of strapped-down astronauts exercising on the Station's treadmills.

The level of microgravity can depend on where you are in the Space Station. Nearer the centre of mass, vibrations are less than compared to the extremities where vibrations are amplified along the long truss. The Space Station's laboratories, including Columbus, are installed around the centre of the outpost for this reason.

Biolab: spinning biology experiments in space

The Biolab rack in Columbus is a complicated machine designed to run biological experiments on micro-organisms, cells, tissue cultures, small plants and small invertebrates. These experiments often aim to identify how weightlessness influences organisms, from the effects on a single cell up to a complex organism including humans.

Biolab works with standard-sized experiment containers around the size of a lunchbox (108 x 150 x 137 mm) that are prepared on Earth. An astronaut inserts the boxes into Biolab where they can be incubated at the right temperature and spun in one of two centrifuges to recreate different levels of gravity. Biolab includes a small glovebox for an astronaut to interact with the experiment.

Some components in Biolab are designed and built by Ferrari, so it is maybe unsurprising that the top-of-the-range performance comes at the cost of high maintenance.

Biolab was the facility used to investigate the time it takes for immune cells from mammals to adapt to microgravity: 42 seconds. Another experiment in Biolab investigated yeast strains that have been used to make bread and brew drinks for centuries. Unsurprisingly, in microgravity the yeast showed signs of stress and had problems building cell walls. The cells diverted their energy to repair themselves and grew less quickly. By analysing the strains that performed better in microgravity, researchers could identify genes that could be used again for longer space missions. Space-faring yeasts can be cultured for future missions to far-away planets – space bakery anyone?

The European Physiology Module

The European Physiology Module in Columbus was developed to study how humans react to spaceflight. It is important to understand how astronauts cope with living in space so we can develop ways to prevent unwanted side effects such as bone and muscle loss. As the human body adapts to living in weightlessness it experiences a form of rapid ageing. Astronauts' eyesight, motor functions, bone and muscle strength and even brains show signs that are typical of elderly people on Earth – except in the body of an exceptionally fit person.

For researchers the astronauts offer a pool of test subjects unlike any they could ever investigate on Earth without getting into trouble with an ethics committee. How else can you put a healthy individual through increased radiation and intense stress and force them to not walk for six months?

Most astronaut investigators collect information before they are launched into space to record a so-called 'baseline reading' and then multiple times during their mission to chart how the body adapts. When the astronauts return to Earth, the tests are then repeated to monitor how the body recuperates to living with gravity's pull.

Cellular science

Studying the effects of spaceflight on the building blocks of life is an opportunity like no other. As cells encounter microgravity they react in many different ways, as if threatened by disease, stress or any unknown invasion. Monitoring how cells behave is both an important and an extremely exciting field of science taking place in space – and discoveries made are already yielding benefits for us back on earth.

The 5-LOX enzyme regulates life expectancy of human cells. Most human cells divide and regenerate, but the number of times they replicate is limited. Italian researchers wanted to find out how this enzyme was affecting astronauts' health in space, so blood samples from two healthy donors were sent to the orbital outpost. One set was exposed to weightlessness for two days, while the other was held in a small centrifuge to simulate Earth-like gravity. The samples were then frozen and sent back to Earth for analysis.

Interestingly, the weightless samples showed more 5-LOX activity than the centrifuged samples, giving scientists a target enzyme that could play a role in the weakening of immune systems. The enzyme can be blocked with existing drugs, so using these findings to improve human health is a close reality. Limiting activity of cell signals such as those controlled by 5-LOX might even slow parts of the ageing process. Who knows what life-prolonging pills could be developed in the future, all thanks to a two-day space experiment.

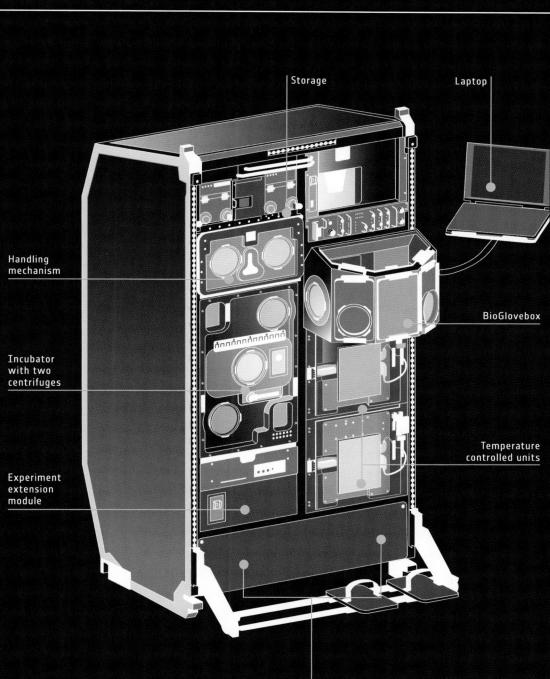

Storage

Laptop

Handling
mechanism

BioGlovebox

Incubator
with two
centrifuges

Temperature
controlled units

Experiment
extension
module

Automatic section   Manual section

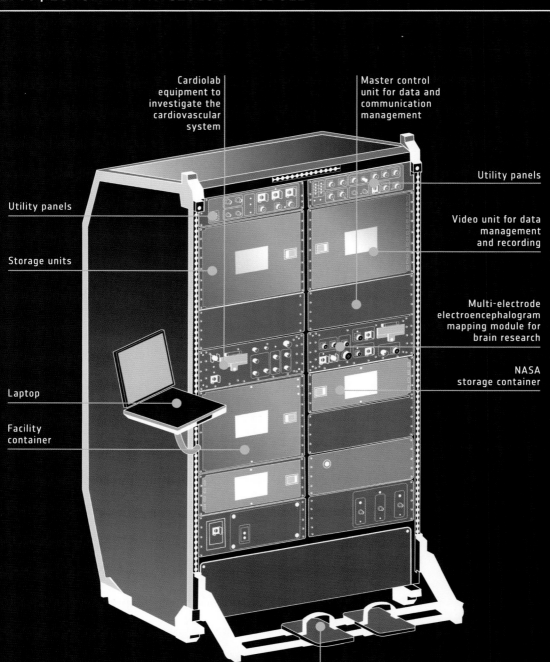

Cardiolab equipment to investigate the cardiovascular system

Master control unit for data and communication management

Utility panels

Video unit for data management and recording

Multi-electrode electroencephalogram mapping module for brain research

NASA storage container

Utility panels

Storage units

Laptop

Facility container

A similar experiment put immune cells (those that fight infection by ingesting microorganisms) through artificial gravity in space and showed that a specific transmitter in the cells, called the Rel/NF-κB pathway, stops working in weightlessness.

Normally, when our bodies sense an invasion, a cascade of reactions occur that are controlled by the information held in our genes. Finding which gene does what is like looking for the right key to fit a keyhole, without having found the keyhole yet.

Studying cells that have flown on the International Space Station is putting researchers on the right path to finding the key to how our immune system works. Comparative samples are showing them where to look to find which genes instruct our immune cells to react to diseases, and how.

Space research could help to tackle disease in two ways. Stopping genes that activate our immune system would help to relieve people suffering from autoimmune diseases such as arthritis. The pharmaceutical industry could also find the genes that need to be active to fight specific illnesses and market tailored antibodies.

Fluids in space to study fluids inside Earth

The Fluid Science Laboratory was designed to allow researchers to investigate liquids in space without molecules being pulled to the bottom of a test tube and therefore influencing the experiment.

The unit uses experiment containers like those used in the Biolab and offers all the necessary equipment to prod and vibrate liquids while observing the result with cameras and lights. As with most Columbus hardware, the design is plug-and-play – once the experiment is inserted it can be run uninterrupted by mission control while the astronauts work on other things.

The Fluid Science Laboratory's flagship experiment, however, is an investigation of how Earth's core behaves.

Space might be the final frontier, but we know more about the Universe above than what is happening under our feet. Just 30 kilometres below the ground we walk on, the thin outer crust, Earth's mantle is a semi-solid fluid. The highly viscous layers there vary with temperature, pressure and depth. The deepest that humans have ever drilled is

FSL | FLUID SCIENCE LABORATORY

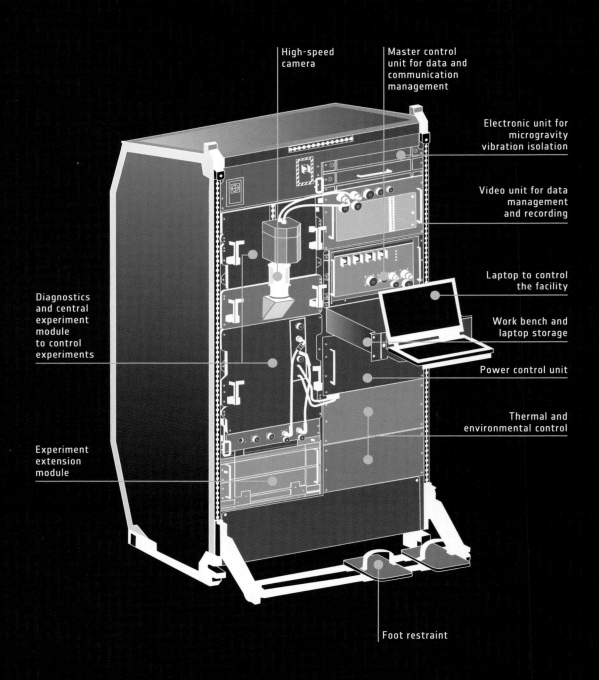

High-speed camera

Master control unit for data and communication management

Electronic unit for microgravity vibration isolation

Video unit for data management and recording

Laptop to control the facility

Work bench and laptop storage

Power control unit

Thermal and environmental control

Diagnostics and central experiment module to control experiments

Experiment extension module

Foot restraint

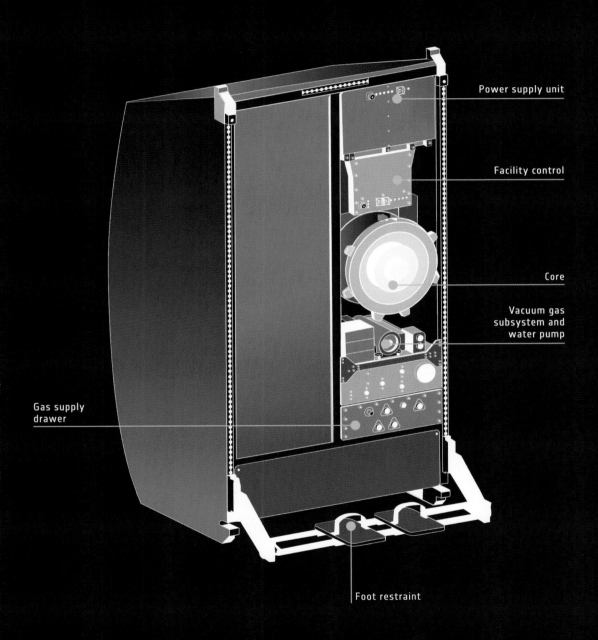

Power supply unit

Facility control

Core

Vacuum gas
subsystem and
water pump

Gas supply
drawer

Foot restraint

just over 12 kilometres, so investigating the depths directly is out of reach, except for in simulations — like one that took place in the Fluid Science Laboratory.

The Geoflow experiment consisted of two revolving concentric spheres with a liquid between them, all housed in a standard experiment container. The inner sphere represented Earth's core, with the outer sphere acting as the crust. The liquid was the mantle, with a high-voltage electrical field creating artificial gravity for the experiment. Of course, Earth's gravity had been removed as a variable.

As the spheres rotated, a temperature difference was created between them. The temperatures could be controlled down to a tenth of a degree.

Plumes of hotter liquid were seen rising towards the outer shell — confirming the predictions of computer simulations about the behaviour of Earth's core. The mushroom-like plumes that were observed could explain geological formations in our planet such as the Hawaiian line of volcanoes in the South Pacific.

The European Modular Cultivation System

Though it predates the laboratory, the European Modular Cultivation System was fitted to Columbus to conduct research into plant growth in microgravity. Growing fresh food will be an important part of missions further afield where steady supplies will not be an option. Food is a huge morale boost and many astronauts dream of a fresh dish after six months of freeze-dried and rehydrated menus. ESA allows one or two special 'bonus' meals for each six-month mission. Created to the astronauts' wishes in collaboration with a three-star chef, the meals are high cuisine — albeit in powdered, dehydrated or canned form.

The European cultivation facility aims to understand how plants grow by looking at root growth and plant cells. Using space-standard containers the system offers everything to incubate plants, from regulating temperature, humidity, oxygen and carbon dioxide to varying the lighting as well as observing the results on video — it even includes two rotors that spin to simulate gravity. Since its launch, great advances have been made in growing plants in space.

One simple-sounding but hardly obvious problem is working out how plants know which way to grow their roots and where they should grow leaves. On Earth, gravity and the Sun would obviously help a seedling decide where to shoot, but what happens in space where none of these aids exist?

The Gravi experiments found that plants can "feel" gravity at a much lower rate than humans can. Batches of plants were put in a spin at different speeds to recreate varying levels of gravity in space. The batches reacted differently to very low gravity levels. Gravity seems to be felt by plants on a cellular level, mainly in the root tips where they use calcium as a signal to grow.

Charles Darwin himself first described how plant stems grow in a corkscrew fashion, but how it happens was unclear. The Multigen experiment in the European mini-greenhouse showed that it is driven by an interplay of light and gravity that sends signals to the plant cells.

Such experiments are not only helping scientists to prepare for far-off colonisation but also improving our knowledge of growing crops at home. If we can work out how plants grow without gravity and in artificial light in space, we can improve crop yields in urban farms, where plants could potentially be stacked three-dimensionally.

The Microgravity Science Glovebox

As Space Station experiment racks are interchangeable they are sometimes moved around. The European-built Microgravity Science Glovebox allows astronauts to work on dangerous experiments in a sealed unit with no risk of the experiment, any potential biological hazards or fire escaping into the Space Station.

By 2012 the glovebox had been used for over 12,000 hours on 24 ESA and NASA experiments, investigating such things as how fire burns in space in order to improve fuel efficiency in satellites and cars, and determining which soluble microparticles could provide the foundations for speed-of-light computers.

Science on the International Space Station is invariably international, as the experiment facilities and research hours are shared between all the partners. A blood sample could be taken in the Japanese Kibo laboratory, analysed in the American Destiny lab and then stored in a freezer in European Columbus. Of the 16 racks in Columbus, four are used by NASA as payment for bringing Columbus to the Space Station on Atlantis. The amount of time each space agency can spend on shared equipment is determined by how much a

EMCS | EUROPEAN MODULAR CULTIVATION SYSTEM

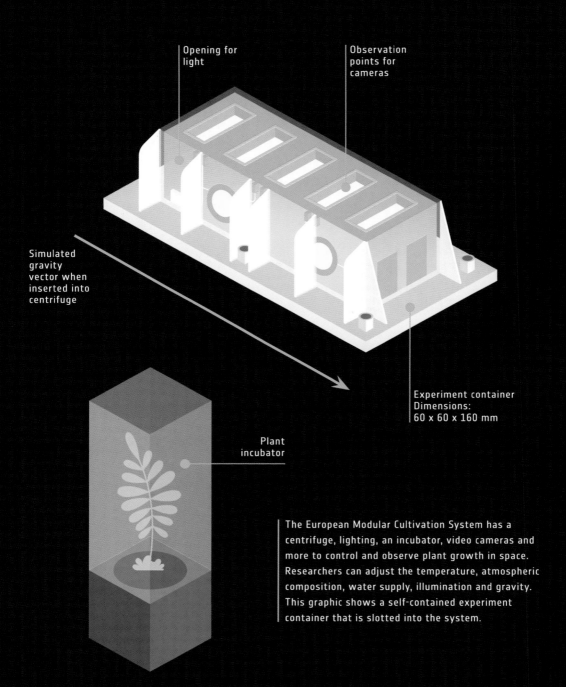

Opening for light

Observation points for cameras

Simulated gravity vector when inserted into centrifuge

Experiment container Dimensions: 60 x 60 x 160 mm

Plant incubator

The European Modular Cultivation System has a centrifuge, lighting, an incubator, video cameras and more to control and observe plant growth in space. Researchers can adjust the temperature, atmospheric composition, water supply, illumination and gravity. This graphic shows a self-contained experiment container that is slotted into the system.

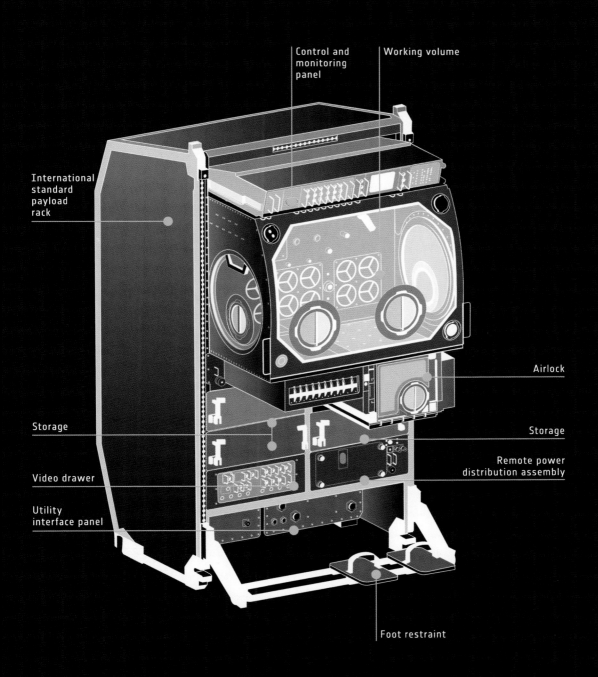

Control and monitoring panel

Working volume

International standard payload rack

Airlock

Storage

Storage

Remote power distribution assembly

Video drawer

Utility interface panel

Foot restraint

space agency contributed to the International Space Station – for ESA this figure is 8.3%.

| Plasma

Plasma is usually a hot, electrically charged gas – the Sun being an extreme example – but "cold plasmas" can also be created at room temperature.

Cold plasmas are an effective way to kill bacteria, fungi, viruses and spores without harming human cells, so they offer a perfect solution for keeping the International Space Station clean and germ-free. Traditional methods to sterilise medical equipment include heating, steaming or boiling instruments until all the bacteria and viruses are killed. This process works fine on Earth but is hardly practical for astronauts on the International Space Station.

Cold-plasma technology was miniaturised and developed to help disinfect the Microgravity Science Glovebox safely, and the technique is now making its way into both hospitals and restaurant kitchens as a quick and easy cleaning solution (it is especially good at keeping the distinctive odour of fried food at bay).

The origin of this plasma innovation was a totally different scientific discipline. The Russian-European Plasma-Kristall experiments were trying to understand how atoms interact and bond to form solid structures such as metals. We know our world is made of atoms and molecules, but even with the most powerful microscopes we cannot see them moving, meaning our understanding of how molecules interact is based on experimentation and assumptions.

Scientists instead recreated larger models of atoms to mimic how they would behave. Statically charged polystyrene in water has been used for decades to understand how atoms (represented by the polystyrene) repel and attract to form a stable structure, but as polystyrene floats on water this technique can only recreate a two-dimensional view. The weightlessness on the International Space Station offered the possibility to investigate in the third dimension by using dust suspended in plasma.

Injecting microscopic dust particles into a neon and argon tube made the dust act as atom substitutes. As the dust floated in the charged gas, it collected negative charges as positive ions accumulated around it. As a result, it started to repulse other, similarly negatively charged, dust particles – just like atoms do in a fluid state.

Doing this research on Earth is not possible — the dust particles would fall with gravity and the simulated atoms would not behave realistically. And so a by-product of trying to understand the atomic world is now allowing students to live above a greasy-spoon without smelling of old oil at university.

Why is there a need to conduct human research? The space environment does not actually offer conditions which humans were made for. There are multiple risks and challenges that astronauts need to face when we send them into space, so if we want to send a human crew to the Moon, Mars or beyond, which is what all major space agencies dream of, there are a lot of problems that we need to solve to ensure that the crew gets to their destination and back safely. The most prominent research platform to use to prepare for such long-duration space exploration missions is the International Space Station and in partnership with the United States, Russia, Japan and Canada, ESA can be proud to participate to the greatest international project ever. The Station has become a shining example of international cooperation, we make excellent use of its capabilities and if we look at the publications stemming from experiments, the European science community is actually at the forefront compared to scientists from the other agencies.

Jennifer Ngo-Anh
ESA head of Space Station human research

The experiment that saves lives

Scientists often call the International Space Station a platform for low-gravity research. The Vessel-ID experiment took this literally and attached an antenna to a platform on the outside of Columbus – a great example of using the Station's full potential to test technology. Its goal was to see if Columbus could capture signals from ships at sea.

Unlike the airline industry, no global marine traffic control exists. Ships broadcast their position and heading via radio. The system works reasonably well, but was designed and implemented before satellites existed, so it only has a reach radius of around 74 kilometres – all ships further than that from the coast are effectively invisible to traffic controllers as well as search and rescue teams.

Radio signals travel in all directions and travel further through space, so Norwegian researchers theorised that even though the radio signals were not designed to be captured by orbiting satellites, a low-flying receiver would still be able to capture the faint signals and process them on a global map.

Launching dedicated satellites to test this theory was daunting – there was a distinct chance that it would not work at all. Instead, Columbus was fitted with a VHF antenna to capture the signals in 2009.

On a good day some 400,000 ship reports are received from more than 22,000 different ships. In October 2011, the total number of position reports exceeded 110 million from more than 82,000 different ships.

The antenna now tracks ships all over the world and helps direct nearby sailors to ships in distress, saving lives by doing so.

With the system proven to work, new generations of receivers on satellites are now offering near-global coverage of the seas, aiding search and rescue, combatting piracy and smuggling and hinting at fuel-saving shortcuts for common shipping routes.

Thermolab

Many good experiments start with a simple question and lead to surprising results. A Berlin-based group of researchers specialising in space medicine and extreme environments wanted to know exactly what happens to an astronaut's body temperature in space.

They proposed a study that would record astronauts' body temperature regularly, but the International Space Station's scientific board warned that the vast amount of temperature readings would not be acceptable – they would cause discomfort to the astronaut and waste their time. Astronauts could not be expected to interrupt work to take their temperature every couple of hours.

Measuring a patient's temperature is a staple part of medicine, diagnostic techniques and human research. From a simple hand on the forehead to a thermometer stuck awkwardly into a patient's mouth to inaudible ultrasound waves blasted into ears, there are many ways to record the body's temperature. Despite some of the techniques having being used for centuries they all have distinct disadvantages: each thermometer records temperature snapshots while causing discomfort; they only reveal local temperature, not core body temperature; and they require an extra pair of hands to execute.

To overcome these downsides, the researchers developed a new type of thermometer called Thermolab. The equipment uses two sensors attached to the forehead and chest and a separate control box to calculate the body temperature.

The sensor works so well that it is used by firefighters and in extreme environments such as Antarctica.

A thermometer that can be read from a distance and continuously records very accurate data has enormous potential. The sensor is already being used during open-heart surgery on children, but also might soon be employed as a low-cost, low-effort general instrument in hospitals to aid the monitoring of patients.

Notably, the experiment showed that an astronaut's temperature rises by 1ºC during the first two months in space and stays there until it drops back to normal on return to Earth.

The Immuno experiment

The immense changes that an astronaut undergoes in space are enough to give anybody stress, and a long-running piece of European research is trying to understand how this affects an astronaut's life. Stress is a response of the body as it adapts to hostile environments. This broad definition includes stress from speaking in front of an audience, stress from a wound or stress from living in weightlessness in a fragile spacecraft far from home.

The Immuno experiment took a holistic approach to investigating stress and requested questionnaires, blood samples and temperature readings from the astronauts.

The researchers saw an ambiguous immune response in the astronauts' blood, over-reacting to some harmless changes while ignoring more dangerous ones.

But while the results from the study might not immediately be of interest for Earth-bound people, the methods used have already benefited patients in hospitals as well as tax payers who contribute to health care.

The first experiment took five years to complete and involved meticulous planning to use every drop of blood taken from the astronauts. Through necessity, the researchers developed new ways of analysing small quantities of blood, so as not to drain the astronauts' already-depleted supplies. Now, the hardware used and the methods are being shared with the medical community to aid the care of at-risk new-borns, who have even less blood to spare for analysis.

Results

The second thing people ask about the International Space Station, after how astronauts go to the toilet, is where the results from all the scientific studies are. The answer to the first question is with a vacuum toilet that sucks all bodily waste into bags. The answer to the second question is that results take time.

Research in space is a continual process, and an extra tool for the vast amount of research being conducted all over the world. It is not simply a given, and is immensely difficult to plan and finance. After it is proven that an experiment can be run only in space, experiments are then measured by how much value they would add to research done already on Earth.

Gone are the days when one scientist would discover or theorise a new phenomenon and put his or her name to it. Scientific advancements are now almost always a collaborative effort, often involving scientists from all over the world. Most people associate the old-fashioned filament light bulb with Thomas Edison and radio with Guglielmo Marconi, but do you know who invented modern LED lights or Wi-Fi networks? These modern versions of lighting and radio communication were developed across borders and even across generations, making it harder to pinpoint specific inventors. Nobel prizes in science are increasingly awarded to teams.

Space-age metals are probably in your pocket or handbag as you read this. All smartphones from a certain leading manufacturer use a scratch-resistant metal alloy that was developed based on research conducted on a Space Shuttle in 1997. It took over 12 years from discovery for this new alloy to find its way into your pocket, so imagine how long it takes for a drug researched in space to be approved after going through clinical trials.

ESA research has already helped develop industrial processes to mass-produce aircraft-grade alloys that are twice as light as conventional nickel alloys while offering equally good properties. Airlines are always looking for ways to save fuel by cutting down on weight without sacrificing safety. Generally, cutting weight by 1% will save up to 1.5% in fuel, while the new titanium aluminide alloy developed will reduce turbine weight by 45% over traditional components. Of course, airlines and safety regulators want to be sure the new metal is reliable over many years of use, so these lighter materials are still being tested before they start propelling people to their holiday destinations.

Space research is still a new field, and discoveries emerge in unexpected ways. Columbus has only been in orbit for 10 years at this point – it will be in the years to come that the findings from these experiments that have taken place in the laboratory pay off for the general public, whether it be in hardier phones, lighter planes, or something entirely different.

We do many experiments in ESA's Columbus laboratory but it also serves as the focal point for life sciences. Astronauts have performed many blood draws, ultrasound examinations, lung evaluations and even electroencephalography (EEG) measurements in Columbus, in addition to many more life science experiments. Columbus even houses a mini-centrifuge, which separates our blood samples into blood plasma and red blood cells before being frozen for later return to Earth – happy times!

Tim Peake
ESA astronaut

10 YEARS OF SCIENCE

ESA EXPERIMENTS RUN ON THE INTERNATIONAL SPACE STATION SINCE THE LAUNCH OF COLUMBUS

Per research discipline

227

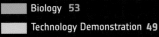 Biology **53**	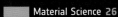 Material Science **26**	 Fundamental Physics **1**
Technology Demonstration **49**	Fluid Physics **14**	Plasma Physics **1**
Human Physiology **41**	Radiation **10**	Earth Observation **1**
Education **28**	Solar Physics **3**	

PEOPLE INVOLVED IN INTERNATIONAL
SPACE STATION EXPERIMENTS FROM COLUMBUS
LAUNCH UNTIL FEBRUARY 2018

Per research discipline

5161

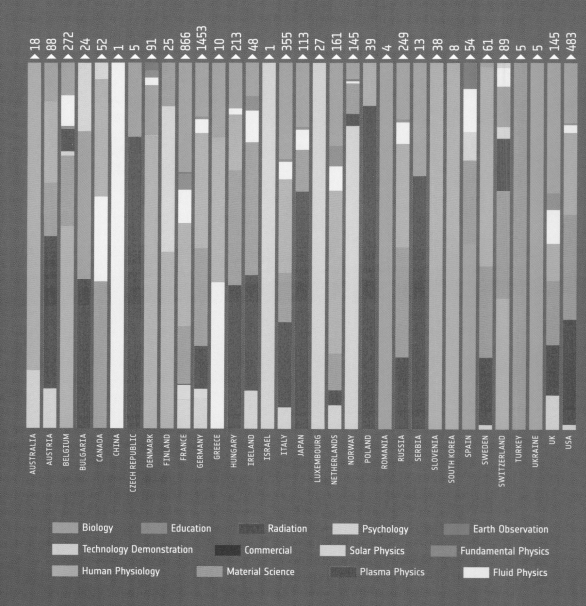

Values per country (top of bars): AUSTRALIA 18, AUSTRIA 88, BELGIUM 272, BULGARIA 24, CANADA 52, CHINA 1, CZECH REPUBLIC 5, DENMARK 91, FINLAND 25, FRANCE 866, GERMANY 1453, GREECE 10, HUNGARY 213, IRELAND 48, ISRAEL 1, ITALY 355, JAPAN 113, LUXEMBOURG 27, NETHERLANDS 161, NORWAY 145, POLAND 39, ROMANIA 4, RUSSIA 249, SERBIA 13, SLOVENIA 38, SOUTH KOREA 8, SPAIN 54, SWEDEN 61, SWITZERLAND 89, TURKEY 5, UKRAINE 5, UK 145, USA 483

Legend:
- Biology
- Education
- Radiation
- Psychology
- Earth Observation
- Technology Demonstration
- Commercial
- Solar Physics
- Fundamental Physics
- Human Physiology
- Material Science
- Plasma Physics
- Fluid Physics

People can be counted twice if they have roles in more than one project

PEOPLE INVOLVED IN INTERNATIONAL SPACE STATION EXPERIMENTS FROM COLUMBUS LAUNCH UNTIL FEBRUARY 2018

Per research discipline

933

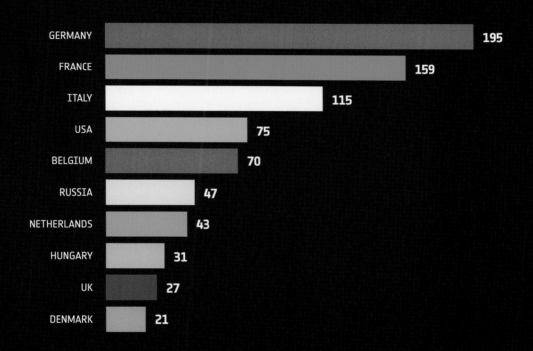

GERMANY	195
FRANCE	159
ITALY	115
USA	75
BELGIUM	70
RUSSIA	47
NETHERLANDS	43
HUNGARY	31
UK	27
DENMARK	21

NORWAY **20**	AUSTRIA **10**	FINLAND **5**	CZECH REPUBLIC **2**	TURKEY **1**
SWITZERLAND **20**	SWEDEN **10**	BULGARIA **3**	ROMANIA **2**	UKRAINE **1**
JAPAN **18**	IRELAND **6**	GREECE **3**	SERBIA **2**	CHINA **1**
SPAIN **17**	AUSTRALIA **5**	LUXEMBOURG **3**	ISRAEL **1**	
CANADA **11**	SLOVENIA **5**	POLAND **3**	SOUTH KOREA **1**	

People are counted once regardless of how many roles or projects they are involved in

EXPERIMENT RUNS IN COLUMBUS MODULE
BY ALL SPACE STATION PARTNERS
SINCE LAUNCH TO SEPTEMBER 2017

Per research discipline

214

Human Research 52

Technology Demonstration 48

Physical Science 47

Biology 38

Education 24

External Payloads 4

Miscellanous 1

EXPERIMENT RUNS IN COLUMBUS MODULE
BY ALL SPACE STATION PARTNERS
SINCE LAUNCH TO SEPTEMBER 2017

By agency

214

ESA 147   NASA 56   CSA 8   ASI 2   JAXA 1

EXPERIMENTS PERFORMED ON THE WHOLE
INTERNATIONAL SPACE STATION BY ALL PARTNERS
FROM COLUMBUS LAUNCH UNTIL SEPTEMBER 2017

Per research discipline

1785

ESA | 219

- 50
- 3
- 28
- 39
- 50
- 49

NASA | 723

- 165
- 95
- 130
- 77
- 64
- 192

CSA | 28

- 2
- 1
- 10
- 9
- 2
- 4

ROSCOSMOS | 397

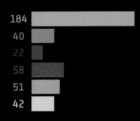

- 184
- 40
- 22
- 58
- 51
- 42

JAXA | 418

- 266
- 17
- 47
- 26
- 18
- 44

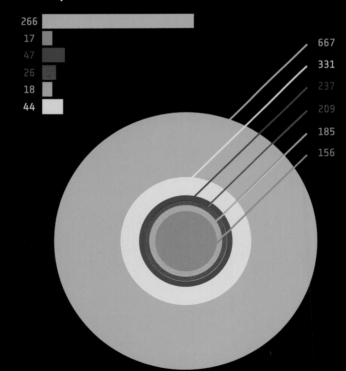

- 667
- 331
- 237
- 209
- 185
- 156

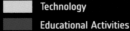

■ Biology and Biotechnology
■ Technology
■ Educational Activities
■ Human Research
■ Physics
■ Earth and Space Science

PARTICIPANTS INVOLVED IN THE WHOLE
INTERNATIONAL SPACE STATION FROM COLUMBUS
LAUNCH UNTIL SEPTEMBER 2017

By agency

3141

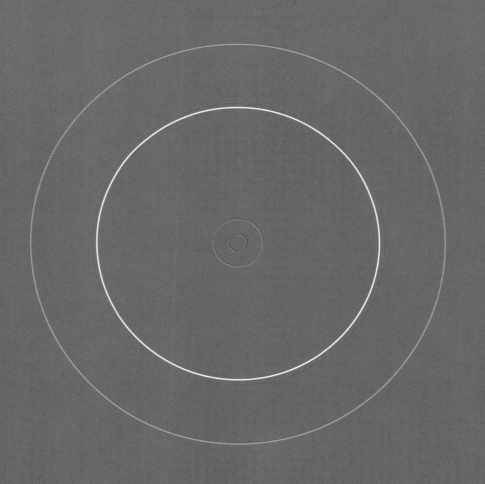

NASA 1233   ESA   JAXA **841**

ROSCOSMOS   CSA 39

EXPERIMENTS PERFORMED ON THE WHOLE INTERNATIONAL SPACE STATION BY ALL PARTNERS FROM COLUMBUS LAUNCH UNTIL SEPTEMBER 2017

Experiments run in cooperation with
multiple space agencies

577 COOPERATIVE
EXPERIMENTS
IN INTERNATIONAL
SPACE STATION

24 COOPERATIVE
EXPERIMENTS
IN COLUMBUS

OPERATED
ON COLUMBUS

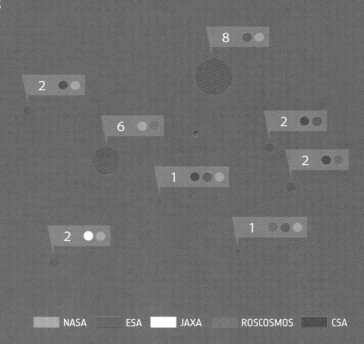

NASA   ESA   JAXA   ROSCOSMOS   CSA

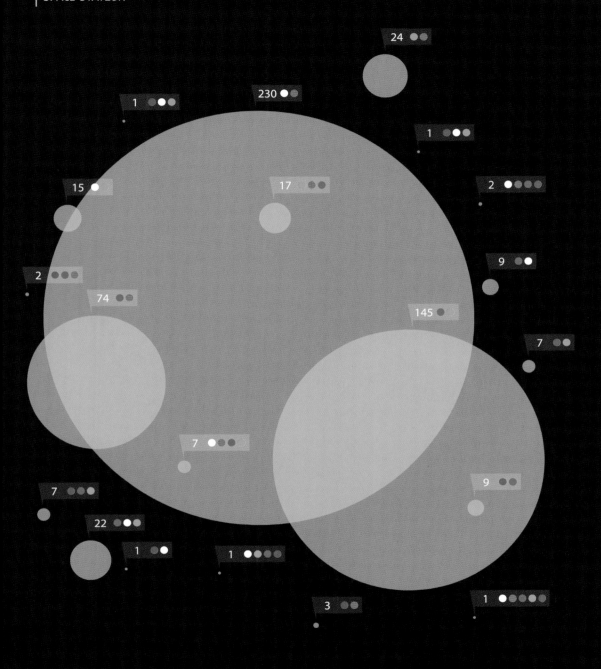

NASA   ESA   JAXA   ROSCOSMOS   CSA

SCIENCE FORECAST

PLANNED EXPERIMENTS TO RUN ON THE WHOLE
INTERNATIONAL SPACE STATION IN 2018

Per research discipline

60

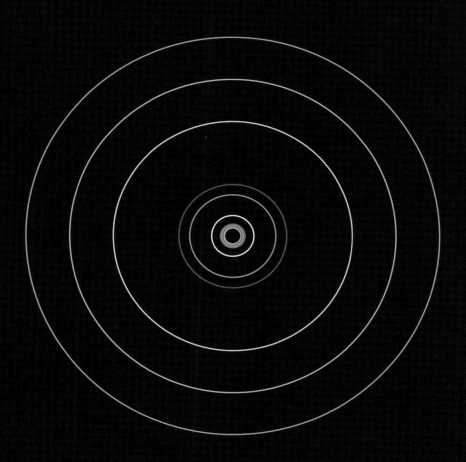

Material Science 19		Fluid Physics **2**	
Human Physiology 15		Biology 1	
Technology Demonstration 11		Radiation 1	
Commercial 5		Earth Observation 1	
Education 4		Plasma Physics 1	

PEOPLE INVOLVED IN
INTERNATIONAL SPACE STATION
EXPERIMENTS FORECAST IN 2018

Per research discipline

689

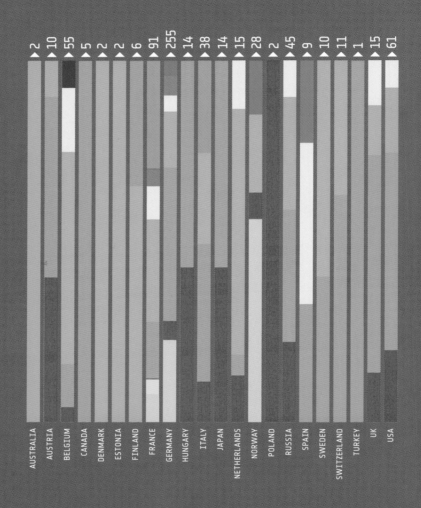

Bar chart values (left to right): AUSTRALIA 2, AUSTRIA 10, BELGIUM 55, CANADA 5, DENMARK 2, ESTONIA 2, FINLAND 6, FRANCE 91, GERMANY 255, HUNGARY 14, ITALY 38, JAPAN 14, NETHERLANDS 15, NORWAY 28, POLAND 2, RUSSIA 45, SPAIN 9, SWEDEN 10, SWITZERLAND 11, TURKEY 1, UK 15, USA 61

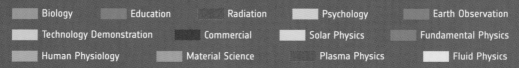

Legend:
Biology · Education · Radiation · Psychology · Earth Observation
Technology Demonstration · Commercial · Solar Physics · Fundamental Physics
Human Physiology · Material Science · Plasma Physics · Fluid Physics

People can be counted twice if they fulfil multiple roles in different projects and can prepare experiments that will not fly until 2019 or later

PEOPLE INVOLVED IN INTERNATIONAL SPACE STATION EXPERIMENTS FORECAST IN 2018

By country

305

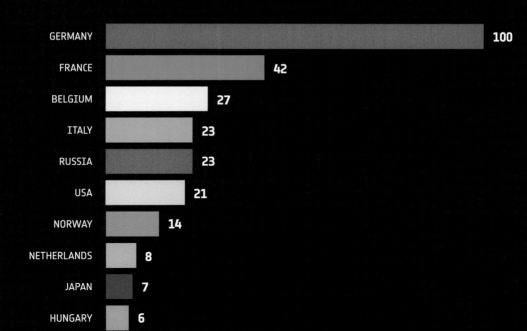

GERMANY	100
FRANCE	42
BELGIUM	27
ITALY	23
RUSSIA	23
USA	21
NORWAY	14
NETHERLANDS	8
JAPAN	7
HUNGARY	6

THE TEAM

Astronauts are many things — engineers, doctors, athletes, linguists — but they cannot be expert scientists for all the experiments they run. Research in space would be pointless without bringing the results to people on Earth who know exactly what to do with them. All results are published in journals available to scientists as part of the various Space Station partnerships, but getting the data and samples home is an achievement in itself.

Transport is limited, so scientists who require samples and hard disks to be returned to Earth need to first consider how long their experiment can be stored while keeping weight and volume to a minimum. Return flights to Earth are limited, so most data is transmitted wirelessly.

The space-to-ground network is an amazing achievement that allows astronauts access to the internet as well as letting them conduct live press conferences, while scientists can operate their experiments and send results at the same time.

Each signal over the communications network in the European laboratory travels from the International Space Station to another satellite some 36,000 kilometres above Earth, through Houston mission control in the USA and across the Atlantic Ocean to one of the European hubs for space-science operations — miraculously only taking up to 0.8 seconds in total both ways.

As the Space Station travels at 28,800 km/h, the time for each signal to reach its destination changes continuously, and there are defined moments when contact is lost altogether for a few minutes when the orbits of the communication satellites and the outpost don't align.

Despite this, the system can download up to 300 million bits per second, twice the rate of high-speed home internet, and the Columbus Control Centre receives 500 gigabytes of data on average per month from space.

Ground control

There are ground controls for the International Space Station all over the world. European researchers can control and observe their experiments at user operations centres in Denmark, Spain, France, Norway, Germany, Belgium, the Netherlands and Switzerland.

COLUMBUS CONTROL CENTRE

Over the years the layout and teams of the Columbus Control Centre has changed, this graphic is a representation of 10 years of operations

1. Columbus data communications

2. COSMO

3. STRATOS

4. Command console for Columbus launch and setup

Live video feed from Space Station A

Columbus operations messages B

Astronauts' schedule C

Space station orbit D

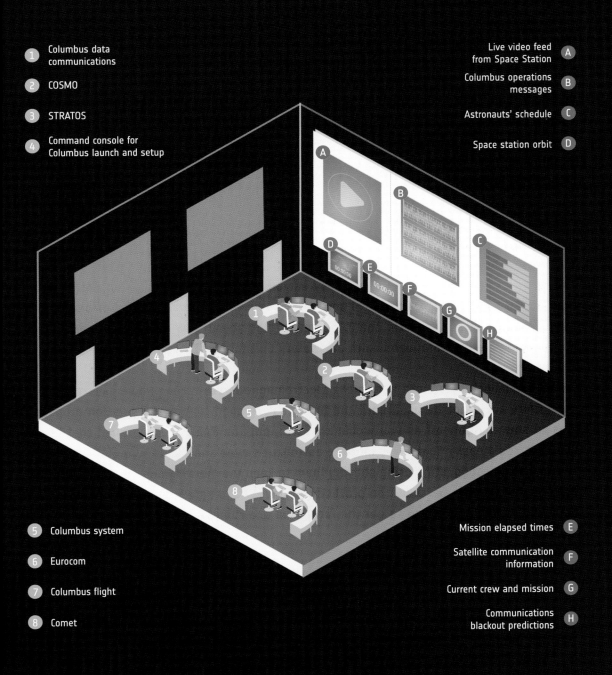

5. Columbus system

6. Eurocom

7. Columbus flight

8. Comet

Mission elapsed times E

Satellite communication information F

Current crew and mission G

Communications blackout predictions H

The Columbus Control Centre, near Munich, Germany, operates 24 hours a day, 365 days a year to ensure the European laboratory stays aloft. A windowless bunker inside a German military base houses the control centre in duplicate. One room can be used if anything goes wrong (as well as allowing for running tests, maintenance or upgrades) while operations continue unaffected in the other.

All data from Columbus is routed through the Columbus Control Centre, from which it is forwarded to the user operations centres for processing and archiving.

Like in a nuclear reactor control room, most of the work that goes on in the Columbus Control Centre is painstaking but requires constant alertness. Any new task for astronauts is planned in detail with all aspects of change considered – it is very much life and death. With this in mind, the teams work in shifts to stay focused. If an alarm sounds they need to react immediately, assess the situation and report solutions as soon as possible.

PASO

The Product Assurance and Safety Officers act as a sort of quality control, assessing every aspect of a task and the objects used by astronauts for safety. Unlike on Earth, small cuts and bruises take longer to heal and chemicals involved linger instead of dissipating harmlessly in the atmosphere. The PASO team think ahead and protect astronauts from harm – they are the European health and safety unit for the International Space Station.

In 2014 a launch bolt broke before the Electromagnetic Levitator could be installed and the facility could not be used until the bolt was removed. The initial solution proposed was to saw the bolt off as you would on Earth, but the PASO team warned that metal shavings could easily float away into the astronaut's eyes and blind them. PASO provided the solution by suggesting using shaving foam to keep the metal filings stuck in place. ESA astronaut Alexander Gerst put on protective eyewear and sprayed shaving cream on the bolt before sawing it in half to get the facility up and running.

COSMO

Ever spent stressful minutes looking for your wallet or keys before you leave the house in a hurry? The Columbus Stowage and Maintenance Officers would have your back if you were an astronaut in space. This team keeps track of the thousands of items that are in the Space Station and can quickly direct astronauts to that elusive misplaced screwdriver they need to install a new experiment — saving time for everyone.

COSMO is involved with other control centre teams when preparing a new task for astronauts. They will supply the information on where objects are located and how long it will take to get them as well as how they should be cleared up afterwards. COSMO frequently thinks outside the box and suggests new applications for objects when needed. For instance, a toothbrush has been taken out on a spacewalk to clean bolts on the outside of the Space Station.

The best part of this job is the variety of activities we do. On any given day we might start by activating a large experiment rack for the day's activities, then perform maintenance on our water pumps and afterwards troubleshoot and recover from a LAN switch unexpectedly turning off... all before most people have had breakfast.

Laura Zanardini
STRATOS

STRATOS

With an abbreviation even the team themselves cannot explain, the Columbus Systems, Data Management and Communications Engineers, dubbed STRATOS, keep Columbus running smoothly. Sitting behind a desk that has more screens and data displays than a financial day-trader (over 20 in total), the STRATOS team observes Columbus's vital signs at all hours of the day.

Part of STRATOS's job is to anticipate alarms — and ignore them if needed. A press conference was held in Columbus shortly after its installation in space. The European laboratory was designed for three astronauts at a time, but the press conference demanded the whole crew of the Space Station and the Space Shuttle to be inside the new module.

Humidity sensors and carbon dioxide alarms shouted out warning STRATOS of impending doom. In this case, the alarms were expected as the full house was a planned activity and deemed safe for a brief period.

"Expected alarms" are part of the job and hardly raise blood pressure for the teams who prepare tasks in advance and know what to expect. If any reading does fluctuate from the norm they will be the first to pick up on the warning signs and offer solutions, often solving problems without astronauts even being aware of them.

EPIC

The European Planning and Increment Coordination teams stay out of the limelight but they do their EPIC acronym proud. Planning for months in advance with parameters that often change, the EPIC team micromanage each astronaut's time to make sure they can work efficiently.

Taking everything into account — from spacecraft schedules to experiment constraints, power distribution limits to radio connections and, of course, astronaut personal preferences — the EPIC team have a lot to process. In the small closed ecosystem of the Space Station each action can have far-reaching consequences, like the proverbial butterfly that flaps its wings and causes a storm on a different hemisphere.

Scheduling a daily run on the Station's treadmill can cause vibrations in the Japanese module where a long-running experiment growing delicate protein crystals could be ruined. Turning on an experiment that toasts plastics as a stand-in for metal alloys could draw too much power away from another experiment that has been running perfectly for days and ruin the whole research.

Judging how long a task will take in space, compared to on Earth, is incredibly hard. Consider how much time a planner should reserve for an astronaut who has to unscrew 96 screws to diagnose a fault, solve the issue and then reinstall the 96 screws.

Generally, the time it takes to complete a task on Earth is increased by 50% in space and doubled during an astronaut's first few weeks in space as they get used to their new surroundings and equipment. When you factor in delays in spacecraft launches, meaning necessary supplies might not be on hand as expected, the team are constantly moving tasks and switching plans right up until the night before they are due to be carried out.

Eurocom

Despite so many teams working together to support astronauts, under normal circumstances, only one group is allowed to talk to them directly: the European capsule communicator, or Eurocom.

Eurocom team members are trained to think and talk like an astronaut and know the Columbus systems equally well. They translate orders and requests into clear-to-understand radio transmissions and formulate replies when an astronaut has a question or problem. Eurocom are the singular voice of the teams from all over Europe, including the science hubs, and make sure the astronauts get clear and concise information at all times.

They also represent the astronauts during meetings to discuss tasks, trying to bring the crew's perspective to any discussion between ground controllers.

The Flight Director

The Flight Director takes final responsibility for Columbus, and is allowed to talk to astronauts and the Space Station partner flight directors directly if needed.

With many teams working on all aspects of Columbus, the Flight Director takes the final decision based on reports from the centre. If anything goes wrong or tough decisions need to be taken, all eyes are on them to make the final call.

I remember the very first Columbus data packets from the Space Station moments after activation and reported: "ISS GC, GSOC GC on IP GC 1: We confirm Columbus telemetry received at Col-CC!" As Ground Controller we speak with many acronyms but the message was clear to everybody at the control centres: Columbus was alive!

Kevin Pasay
Ground controller during Columbus installation mission

THE FUTURE OF EXPLORATION

Like Antarctic explorers and alpinists of old, the first space explorers went on daring one-off missions to test the mystery of space. Explorers of the unknown returned with new maps and catalogued the experience for future missions, allowing the unexplored to become less intimidating. Bases were built for scientific research, and living and working in an inhospitable environment became somewhat routine.

The International Space Station can be seen as a first outpost and research base in a place that until 60 years ago was the realm of deities, science fiction and imagination.

The research and work done in space will benefit future generations in ways we can hardly imagine, but it is up to the future generations to continue the effort. Fittingly, astronauts and spaceflight are a well-known source of inspiration for youngsters and educators alike.

In the Netherlands the "André Kuipers" effect is named after the first Dutch ESA astronaut to spend six months on the International Space Station. Statistics show a significant increase of applications to scientific studies 10 years after his first flight in 2004 — a generation of youngsters seem inspired to follow the floating feet of their astronaut hero.

Proving what is possible and improving operations

Working inside the International Space Station is sometimes like assembling complex furniture, but with the tools and instructions continually floating out of reach. The steps required to install complicated equipment or maintain the Station's intricate life-support system can never be memorised, so astronauts follow step-by-step instructions on laptops or tablets.

This is simple and secure but astronauts need to have at least one hand free to move to the next step and a lot of time is lost referring to procedures and returning to the task at hand. To improve on this, ESA is looking at heads-up displays, voice commands and over-the-shoulder cameras to free the astronauts' hands and time — using the International Space Station as a test-bed for designs.

You might think these technologies are already readily available — why don't astronauts have them yet? When working on equipment with your life potentially at risk, your tools must be faultless — they can't misinterpret a voice command, for example. There is little room for error, and as we venture further from Earth the error margin gets smaller.

The International Space Station is a test-bed and proving ground for exploration. Only by mastering living in low Earth-orbit will we be able to explore further afield. The first International Space Station expedition was launched in November 2000 and ever since there have been at least two humans above our heads circling our planet.

Along the way we have learnt a lot about operating a spaceship and living in weightlessness, improving incrementally over the years. For example the first astronauts to live on the Space Station mainly drank water that was shipped from Earth, by 2009 they were drinking their own recycled urine, and now it is practically a closed loop requiring no new water to be shipped from Earth.

Using Earth as a stand-in for Mars

The International Space Station offers a great platform for testing the communications technology for future exploration of planets other than Earth.

Landing humans on a distant object is one thing, but they will also need the fuel and equipment to return home when done. Sending robots to scout landing sites and prepare habitats for humans is more efficient and safer, especially if the robots are remotely controlled by astronauts, who can react and adapt to situations better than a computer.

Radio signals take up to 12 minutes to reach our nearest neighbour, Mars, so it could take 24 minutes before an operator would know how a robot reacted to a new command. To train for this feedback delay, ESA is letting astronauts control robots on Earth as they orbit our planet.

What started with controlling LEGO-built prototypes to test the network rapidly evolved into controlling robots to insert pins into holes with millimetre precision. The Meteron project has allowed for an astronaut to conduct a remote handshake from space to Earth, all while orbiting at speeds of 28,800 km/h.

The system's adaptability and robust design means it can be used over normal data cellphone networks. This makes it well suited for remote areas that are difficult or dangerous for humans to access, such as offshore drilling sites or nuclear reactors, or where natural disasters have destroyed other communication networks – your cellphone connection is faster than the International Space Station's.

> *Since 2012 the Columbus module has been the driver's seat for a variety of rovers. Several astronauts have been preparing for future robotic missions to the Moon and Mars by remote-driving real rovers on the surface of a heavenly body from an orbiter. Each of these international collaboration experiments has given us essential insight into how future robotic missions should be set up when it comes to communication, robotics and operations. Learning by doing. More such experiments with astronauts in the Columbus module are being prepared. This is one of the key reasons the International Space Station is there, and Columbus is a great European rover operations test facility!*
>
> Kim Nergaard
> METERON Head of the Advanced Mission Concepts Section

To the Moon and beyond

One important aspect of the International Space Station is simply proving that when we work together, we can do everything we set our minds to. Countries from all over the globe – north and south, east and west – worked together for the first time to create a peaceful laboratory for research. Inside the International Space Station, Columbus is another example of international collaboration, involving nine European countries collaborating beyond traditional political alliances. The European Space Agency has members that are not part of the European Union – even welcoming Canada as contributing state.

The partners that run the Space Station are now looking towards continuing the collaboration to build a refuge even further afield, in orbit between the Moon and Earth. This outpost will be used as a gateway to the Moon and one day, hopefully, Mars.

Like a mountain refuge, the deep space gateway will offer shelter and a place to stock up on supplies for astronauts en route to further destinations. It will also offer a place to relay communications and can act as a base for scientific research. From there, a research station on the Moon is not far off; a telescope on the lunar far side would allow astronomers to peer further into the Universe than ever before, using the Moon as a natural 3,474-km-thick shield from humanity's electromagnetic noise – generated by millions of televisions and radios that drown out faint signals from the edge of the Universe.

Unlike previous one-off missions to the Moon, the explorers of the future will build a more sustainable base using resources onsite instead of taking everything from home – using equipment and technologies that have been developed and tested in the International Space Station and the European Columbus laboratory.

Thanks to incremental advances on the International Space Station, space agencies now ship fewer supplies as more oxygen and water is recycled and reused. Research with plants in Columbus has helped understand how we can grow the food to eat on longer missions – bringing the dream of a spacecraft that can support astronauts completely self-sufficiently tantalisingly close to reality.

Experience gained in daily operations by mission control is influencing plans for missions where astronauts will be even further away. Improvements from how we train astronauts to how we remind them of task planning, scheduling and technological aids, are all being made thanks to feedback from ESA astronauts.

By studying these astronauts, researchers are developing countermeasures to the pressures of living in space for longer periods of time. These will ensure that if we ever land on Mars, humans will have the physical strength to walk away after a six-month interplanetary flight. From compact exercise machines to "skinsuits" that compress the spine and a new class of headaches that only occur in space – these problems and solutions can only really be worked through in space – on the International Space Station.

The real test to overcome any difficulties is by doing, and the Columbus laboratory has been offering a space for out-of-this-world research for over 10 years in one of the harshest environments humans can survive.

Thousands of people work and have worked to make this stepping stone viable, all driven by an immense curiosity and a drive to test the limits of what is possible. With results still coming in and Columbus in middle age we can be sure of only one thing going forward: this is just the beginning of humanity's adventure.

I feel that ESA made a very big step forward with delivery and operations of this fabulous space infrastructures and being part of it has been just such an amazing personal endeavour: for the very first time in human spaceflight, we have shown that we cannot only develop amazing spaceflight hardware, but also operate it for a long period of time and assume end-to-end responsibility throughout all the project phases, including decommissioning (which is hopefully still a good while down the road). We have gained – and still continue to do – an immense amount of operations expertise over the past 10 years, but I feel that there is still so much to learn.

Hans Bolender
ESA astronaut training division